SpringerBriefs in Physics

SpringerBriefs in Physics are a series of slim high-quality publications encompassing the entire spectrum of physics. Manuscripts for SpringerBriefs in Physics will be evaluated by Springer and by members of the Editorial Board. Proposals and other communication should be sent to your Publishing Editors at Springer.

Featuring compact volumes of 50 to 125 pages (approximately 20,000–45,000 words), Briefs are shorter than a conventional book but longer than a journal article. Thus Briefs serve as timely, concise tools for students, researchers, and professionals.

Typical texts for publication might include:

- A snapshot review of the current state of a hot or emerging field
- A concise introduction to core concepts that students must understand in order to make independent contributions
- An extended research report giving more details and discussion than is possible in a conventional journal article
- A manual describing underlying principles and best practices for an experimental technique
- An essay exploring new ideas within physics, related philosophical issues, or broader topics such as science and society

Briefs are characterized by fast, global electronic dissemination, straightforward publishing agreements, easy-to-use manuscript preparation and formatting guidelines, and expedited production schedules. We aim for publication 8–12 weeks after acceptance.

More information about this series at http://www.springer.com/series/8902

José Antonio Oller

A Brief Introduction to Dispersion Relations

With Modern Applications

 Springer

José Antonio Oller
Departamento de Física
Universidad de Murcia
Murcia, Spain

ISSN 2191-5423   ISSN 2191-5431   (electronic)
SpringerBriefs in Physics
ISBN 978-3-030-13581-2   ISBN 978-3-030-13582-9   (eBook)
https://doi.org/10.1007/978-3-030-13582-9

Library of Congress Control Number: 2019932717

This Springer imprint is published by the registered company Springer Nature Switzerland AG
The registered company address is: Gewerbestrasse 11, 6330 Cham, Switzerland

Contents

Introduction

Abstract This is an introductory text on the vast and rich field of dispersion relations in hadron physics, with some developments as well of direct interest in nuclear physics. We typically consider the relativistic dynamics, but some chapters are dedicated to the problem of nonrelativistic scattering. The latter has the pedagogical advantage of being a framework with a setup that can be stated in a less abstract way than its relativistic counterpart. Needless to say that the basic principles of dispersion relations, namely, analyticity and unitarity, overpass the standard fields of application of hadron and nuclear physics and, therefore, the exposition here might be also of interest for a wider audience. In this regard, we devote the first four chapters of the book on the basic aspects of the dispersion relations for the scattering amplitudes and related process like production ones. Special emphasis is given to analyticity, unitarity, partial-wave expansion of dynamical amplitudes and the important mathematical theorem of Sugawara–Kanazawa. The latter addresses the question of how many subtraction constants are needed in a dispersion relation. Its demonstration also illustrates a method to determine the asymptotic behavior of dispersive integrals. These results are exemplified in Chap. 5 by developing the exact dispersion relations for the eigenvalues of the scattering kernel in nonrelativistic scattering. As an outcome, it also gives the requirement for the convergence of the Born series in potential scattering. There is then the set of Chaps. 6–12, which deal with the application of dispersion relations to study scattering in partial waves. In these chapters, resonance scattering is also considered, and the examples of the σ or $f_0(500)$ and $\rho(770)$ are discussed in detail. There are several methods exposed to work out the partial-wave amplitudes, depending on how the crossed-channel discontinuities required in the resulting unitarized non-perturbative expression are implemented. Chapters 13–15 are dedicated to the problem of final-state interactions in production processes (similar techniques could be used for initial-state interactions or for implementing both simultaneously, if required). The exposition in this regard begins by determining the unitarity requirements, both in the uncoupled and coupled cases. The former case is treated in detail in Chap. 14 when analyzing the Omnès solution. We discuss a possible problem that could arise in this case by not taking properly into account the presence of zeroes and poles in the Omnès function. This can happen due to a transition from a region of parameters of the model into another one, which could a pathological behavior in the Omnès function. An explicit example is worked out. In Chap. 15, we treat the coupled channel version of the problem and discuss the Muskhelishvili–Omnès problem. There is a pedagogical discussion about the number of independent acceptable solutions to this problem, which is not easy to find in the literature. Chapter 16 is dedicated to the near-threshold non-relativistic scattering, and a comparison is given between general results derived from analyticity and unitarity, with some specific models in potential scattering. Finally, in the last chapter, we exemplify the application of dispersion relations in the nuclear medium by evaluating the contribution of the in-medium nucleon–nucleon interactions to the energy density in nuclear matter. The appendix provides a numerical method for numerically evaluating dispersive integrals. This is particularly interesting when the dispersion relations give rise to a set of integral equations that must be solved to obtain the desired response.

Murcia, Spain   José Antonio Oller

Chapter 1
S and T Matrices. Unitarity

The typical situation of a scattering process that we deal in the subsequent is that corresponding to short-range interactions. Therefore, in the asymptotic past and future the initial and final states of particles, respectively, behave as free ones. The corresponding states are given by the direct product of monoparticle states, each of them being characterized by its three-momentum $\mathbf{p}$, spin s, third component of spin σ, mass m and other quantum numbers (like charges) are denoted globally by λ. The corresponding state is written as

$$|\mathbf{p}, \sigma, m, s, \lambda\rangle, \tag{1.1}$$

with $\sigma = -s, -s + 1, \ldots, s - 1, s$. These states have the relativistic invariant normalization

$$\langle \mathbf{p}', \sigma', m', s', \lambda' | \mathbf{p}, \sigma, m, s, \lambda \rangle = \delta_{s's}\delta_{\sigma'\sigma}\delta_{\lambda'\lambda}(2\pi)^3 2p^0\delta(\mathbf{p}' - \mathbf{p}), \tag{1.2}$$

where $p^0 = \sqrt{m^2 + \mathbf{p}^2}$ is the on-shell energy.

The probability amplitude for an initial state $|i\rangle$ at time $t \to -\infty$ to evolve into a final state $|f\rangle$ at time $t \to +\infty$ is given by the matrix elements of a unitary operator S denoted as the S matrix [1],

$$SS^\dagger = S^\dagger S = I. \tag{1.3}$$

Its matrix elements S_{fi} correspond to

$$S_{fi} = \langle f|S|i\rangle. \tag{1.4}$$

Because of the space-time homogeneity these matrix elements are always accompanied by an energy-momentum Dirac delta function, $(2\pi)^4\delta^{(4)}(p_f - p_i)$, where p_i and p_f are the initial an final four-momenta, in order. In linear relations a Dirac delta function of total energy and momentum conservation factors out while, in nonlinear

J. A. Oller, *A Brief Introduction to Dispersion Relations*, SpringerBriefs in Physics,
https://doi.org/10.1007/978-3-030-13582-9_1

relations (like unitarity), others remain, multiplying S-matrix elements of clusters of particles which control the momentum loops in the processes. We do not always show explicitly the cancelation of the total energy-momentum Dirac delta functions, though the context makes it clear.

In the Dirac or interacting picture of Quantum Field Theory (QFT), with $\mathcal{L}_{int}$ the interacting Lagrangian, the S-matrix is given by the evaluation of the matrix elements

$$S_{fi} = \frac{\langle f | e^{i \int d^4 x \mathcal{L}_{int}} | i \rangle}{\langle 0 | e^{i \int d^4 x \mathcal{L}_{int}} | 0 \rangle}, \tag{1.5}$$

with $|0\rangle$ the free state without any particle (or 0_{th}-order perturbative vacuum). In the previous equation $U(+\infty, -\infty) = \exp i \int d^4 x \mathcal{L}_{int}(x)$ is the evolution operator in the interacting picture from/to asymptotic times and, therefore, S_{fi} is its matrix element between the pertinent final and initial states. The denominator is a normalization factor that removes the disconnected contributions without involving any external particle.

Associated with the S matrix we also have the T matrix, which at least requires the presence of one interaction. Its relation with the S matrix is

$$S = I + iT. \tag{1.6}$$

In terms of the T matrix the unitarity relation of Eq. (1.3) reads

$$T - T^\dagger = iTT^\dagger \tag{1.7}$$

$$= iT^\dagger T, \tag{1.8}$$

by using either the first term or the second one from left to right in Eq. (1.3), respectively. By including a resolution of the identity between the product of two T matrices, we have for the matrix elements

$$\langle f | T | i \rangle - \langle f | T^\dagger | i \rangle = i \sum \int \left[(2\pi)^4 \delta^{(4)} (p_f - \sum_{i=1}^{n} q_i) \prod_{i=1}^{n} \frac{d^3 q_i}{(2\pi)^3 2q_i^0} \right] \tag{1.9}$$
$$\times \langle f | T^\dagger | \mathbf{q}_1, \sigma_1, m_1, s_1, \lambda_1; \ldots; \mathbf{q}_n, \sigma_n, m_n, \lambda_n \rangle$$
$$\times \langle \mathbf{q}_1, \sigma_1, m_1, \lambda_1; \ldots; \mathbf{q}_n, \sigma_n, m_n, \lambda_n | T | i \rangle,$$

where the total energy-momentum conservation, $p_f = p_i$, should be understood. We also have the similar term in the right-hand side (rhs) but with $T^\dagger$ and T exchanged. The sum extends over all the possible intermediate states allowed by the appropriate quantum numbers and with thresholds below the total center-of-mass (CM) energy $\sqrt{p_f^2}$ (otherwise the intermediate Dirac delta function would vanish).

The basic content of Hermitian unitarity (chapter 4.6 of Ref. [2]) is precisely to show that the matrix elements of $T^\dagger$ are also given by the same analytical function as those of T itself but with a slightly negative imaginary part in the total energy (or partial ones for subprocesses) along the real axis, instead of the slightly positive imaginary part used for the matrix elements of T in Eq. (1.9). Therefore, the unitarity relation of Eq. (1.9) gives rise to the presence of the right-hand cut (RHC) or unitarity cut in the scattering amplitudes for the total energy real and larger than the smallest threshold, typically a two-body state. It also embraces other singularities like pole ones, while its iteration from the simplest singularities (pole and normal thresholds) generates more complicated ones such as the anomalous thresholds (sections 4.10 and 4.11 of Ref. [2]).

The factor between square brackets on the rhs of Eq. (1.9) is the differential phase space of the intermediate state $|\mathbf{q}_1, \sigma_1, m_1, \lambda_1; \ldots; \mathbf{q}_n, \sigma_n, m_n, \lambda_n\rangle$. We designate it by dQ and it is worth writing it isolatedly, given its importance in collision theory,

$$\int dQ = \int (2\pi)^4 \delta^{(4)}(p_f - \sum_{i=1}^{n} q_i) \prod_{i=1}^{n} \frac{d^3 q_i}{(2\pi)^3 2q_i^0}. \tag{1.10}$$

Notice that the phase factor is Lorentz invariant.

In general the final and the initial states do not need to contain the same particles, even in nonrelativistic scattering. The latter is a valid limit as long as the three momenta of the particles involved are much smaller than their masses. This condition is required because then the Compton wavelength is much smaller than the De Broglie wavelength, $\hbar/mc \ll \hbar/|\mathbf{p}|$, and we can consider that measuring position is meaningful within good accuracy [3].

Given an initial state of two particles with four-momenta p_1 and p_2, its cross section to a final state $|f\rangle$, denoted by σ_{fi}, is defined as the number of particles scattered per unit time divided by the incident flux ϕ_0. The latter division is necessary because the number of collisions rises in a given experiment as the number of incident particles. In our normalization, Eq. (1.2), we have the following expression for σ_{fi} in the CM,

$$\sigma_{fi} = \frac{1}{4|\mathbf{p}_1|\sqrt{s}} \int dQ_f \, |\langle f|T|\mathbf{p}_1, \sigma_1, m_1, s_1, \lambda_1; \mathbf{p}_2, \sigma_2, m_2, s_2, \lambda_2\rangle|^2, \tag{1.11}$$

where s is the Lorentz invariant $s = (p_1 + p_2)^2$.

For pedagogical reasons we explain how the different factors arise in the previous formula. First, we take the modulus squared of the matrix element of the T matrix, from where a factor $[(2\pi)^4\delta^{(4)}(p_f - p_i)]^2$ arises. One of this Dirac delta function is included in the phase space dQ_f, while the other gives rise to the diverging factor $V\mathcal{T}$, with V the volume of space and $\mathcal{T}$ the interaction time. The latter cancels because we have to divide by the time of interaction $\mathcal{T}$, since we are seeking the transition probability per unit time. On the other hand, the number of states corresponding to the normalization of monoparticle states, Eq. (1.2), is $V2p^0$. Therefore, the flux factor $\phi_0 = 4p_1^0 p_2^0 v_{rel} V$, which takes into account that there are $4p_1^0 p_2^0 V^2$ interacting particles with a relative velocity v_{rel}. The latter is given in the CM ($\mathbf{p}_1 + \mathbf{p}_2 = 0$) by

$$v_{rel} = \left| \frac{\mathbf{p}_1}{p_1^0} - \frac{\mathbf{p}_2}{p_2^0} \right| = \frac{|\mathbf{p}_1|(p_1^0 + p_2^0)}{p_1^0 p_2^0}. \tag{1.12}$$

Notice that in the CM $p_1^0 + p_2^0 = \sqrt{s}$. As a result the factors of V cancel in the calculation of σ_{fi} and we are left with Eq. (1.11). In particular, the total cross section from the initial state, σ_i, is given by the sum over all the possible final sates. From Eq. (1.11) we then have

$$\sigma_i = \frac{1}{4|\mathbf{p}_1|\sqrt{s}} \sum_f \int dQ_f \, |\langle f|T|\mathbf{p}_1, \sigma_1, m_1, s_1, \lambda_1; \mathbf{p}_2, \sigma_2, m_2, s_2, \lambda_2\rangle|^2. \tag{1.13}$$

Needless to say that the sum over f could also involve continuous variables and then instead of a discrete sum (as symbolically indicated in the previous equation) one would have to perform the corresponding integrals.

In the following, for brevity in the notation, we designate the monoparticle states by $|\mathbf{p}_1\sigma_1\lambda_1\rangle$, omitting some labels that might be inferred from the information given.

We can relate the total cross section σ_i with the imaginary part of the forward T-matrix element T_{ii} by taking $|f\rangle = |i\rangle$ in the unitarity relation of Eq. (1.9). We then have

$$\Im T_{ii} = \frac{1}{2} \sum_f \int dQ_f |T_{fi}|^2 = 2|\mathbf{p}_1|\sqrt{s}\,\sigma_i. \tag{1.14}$$

This result is usually referred as the optical theorem.

Had we taken instead the other order $TT^\dagger$ in the unitarity relation then we have

$$\Im T_{ii} = \frac{1}{2} \sum_f \int dQ_f |T_{if}|^2. \tag{1.15}$$

Comparing with Eq. (1.14) we then have the reciprocity relation

$$\sum_f \int dQ_f |T_{fi}|^2 = \sum_f \int dQ_f |T_{if}|^2. \tag{1.16}$$

As a consequence one could derive the important Boltzmann H-theorem in statistical mechanics (chapter 3.6. of Ref. [4]). Let P_i be the probability distribution of having a state i in an infinitesimal phase-space volume around this state, then its variation in time is governed by the balance of states f ending in i and the evolution from i to any other state f. Thus,

$$\frac{dP_i}{dt} = \sum_f \int dQ_f |T_{if}|^2 P_f - P_i \sum_f \int dQ_f |T_{fi}|^2. \tag{1.17}$$

By summing over all initial state i it is clear that

$$\sum_i \int \frac{d^4 p_i}{(2\pi)^4} \int dQ_i \frac{dP_i}{dt} = 0, \tag{1.18}$$

where the first integration involves the total four-momentum of the state i, to remove the extra factor of $(2\pi)^4 \delta^{(4)}(p_i - \sum_j q_j)$ included in dQ_i in the next integral symbol. Physically it represents to allow all possible CM motion, since we are summing over all state i.

In order to simplify the derivation of the Boltzmann theorem, let us take the discretized version of the probability distribution function. Then, the entropy is defined by $S = -\sum_i P_i \log P_i$ (there would be just a constant of difference with respect to taking the continuum distribution probability function) and its derivative with respect to time is

$$\frac{dS}{dt} = -\sum_i (\log P_i + 1) \frac{dP_i}{dt} = -\sum_i \frac{dP_i}{dt} - \sum_i \frac{dP_i}{dt} \log P_i. \tag{1.19}$$

The term $-\sum_i dP_i/dt = 0$ because of Eq. (1.18), while for the last term we use the balance Eq. (1.17)

$$\frac{dS}{dt} = -\sum_i \frac{dP_i}{dt} \log P_i = -\sum_{i,j} \log P_i \left(P_j |T_{ij}^D|^2 - P_i |T_{ji}^D|^2 \right), \tag{1.20}$$

with the superscript D indicating that the modulus squared of the matrix element contains the factor $(2\pi)^4 \delta^{(4)}(p_j - p_i)$, which is symmetric under $i \leftrightarrow j$. Exchanging the indices i and j in the last term of Eq. (1.20), we are left with

$$\frac{dS}{dt} = \sum_{i,j} |T_{ij}^D|^2 P_j \log \frac{P_j}{P_i}. \tag{1.21}$$

Now one makes use of the inequality for any two positive quantities P_i and P_j, $P_j \log(P_j/P_i) \geq P_j - P_i$.[1] Then, the rhs of Eq. (1.21) is larger or equal than

---

[1] For $P_j \geq P_i$ this is clear because then $\log P_j/P_i \geq 1$. In the range $P_j \in [0, P_i]$ the difference $P_j \log(P_j/P_i) - P_j + P_i$ is ≥ 0, because it has a negative derivative with respect to P_j and it is zero at $P_j = P_i$ (it is P_i for $P_j = 0$).

$\sum_{i,j} |T_{ij}^D|^2 (P_j - P_i)$. Exchanging again the indices i and j in the last term we are then left with the inequality

$$\frac{dS}{dt} \geq \sum_{i,j} P_j \left(|T_{ij}^D|^2 - |T_{ji}^D|^2 \right) = 0, \tag{1.22}$$

where in the last step we have taken into account the unitarity implication of Eq. (1.18).

Chapter 2
Two-Body Scattering. Partial-Wave Expansion

We focus mainly on the scattering between two particles of four-momenta p_1 and p_2 going into a final state of two other particles of four-momenta p_3 and p_4. The types of particles in the final and initial states might be different. For a two-body final state the differential phase-space factor of Eq. (1.10) expressed with variables in the CM is

$$dQ = \int \frac{d^3 p_1}{(2\pi)^3 2 p_1^0} \frac{d^3 p_2}{(2\pi)^3 2 p_2^0} (2\pi)^4 \delta(p_1 + p_2) = \frac{|\mathbf{p}_1| d\Omega}{16\pi^2 \sqrt{s}}, \qquad (2.1)$$

where $d\Omega$ is the differential of solid angle of $\mathbf{p}_1$ in the CM. Here we have also introduced the Mandelstam variable s,

$$s = (p_1 + p_2)^2 = (p_3 + p_4)^2, \qquad (2.2)$$

which is equal to the total energy squared in the CM [it was alluded above in connection with Eq. (1.11)]. The other standard Mandelstam variables t and u are defined as

$$t = (p_1 - p_3)^2 = (p_2 - p_4)^2, \qquad (2.3)$$
$$u = (p_1 - p_4)^2 = (p_2 - p_3)^2.$$

From Eqs. (2.2) and (2.3) it follows that

$$s + t + u = m_1^2 + m_2^2 + m_3^2 + m_4^2. \qquad (2.4)$$

In terms of dQ the unitarity relation of Eq. (1.9) below the threshold of states with three or more particles reads

© The Author(s), under exclusive licence to Springer Nature Switzerland AG 2019
J. A. Oller, *A Brief Introduction to Dispersion Relations*, SpringerBriefs in Physics,
https://doi.org/10.1007/978-3-030-13582-9_2

$$\langle f|T|i\rangle - \langle f|T^\dagger|i\rangle = i \sum \int \frac{|\mathbf{p}_1|d\Omega}{16\pi^2\sqrt{s}} \langle f|T^\dagger|\mathbf{q}_1, \sigma_1, m_1, s_1, \lambda_1; \mathbf{q}_2, \sigma_2, m_2, \lambda_2\rangle$$
$$\times \langle \mathbf{q}_1, \sigma_1, m_1, \lambda_1; \mathbf{q}_2, \sigma_2, m_2, \lambda_2|T|i\rangle. \tag{2.5}$$

The differential cross section between two-body states is, cf. Eq. (1.11),

$$\frac{d\sigma}{d\Omega} = \frac{|\mathbf{p}_1'|}{|\mathbf{p}_1|} \frac{|\langle \mathbf{p}_1', \sigma_1', \lambda_1'; \mathbf{p}_2', \sigma_2', \lambda_2'|T|\mathbf{p}_1, \sigma_1, \lambda_1; \mathbf{p}_2, \sigma_2, \lambda_2\rangle|^2}{64\pi^2 s}. \tag{2.6}$$

We have employed here that because of energy conservation $p_1^0 + p_2^0 = p_1'^0 + p_2'^0$.

Rotational symmetry implies the invariance of the S- and T-matrix operators under any rotation R,

$$RSR^\dagger = S, \tag{2.7}$$

$$RTR^\dagger = T. \tag{2.8}$$

In the manipulation that follow we only show the active variables characterizing a state and suppress any other label. If the particles involved in the scattering process have zero spin, then we have that the matrix elements of T are scalars because

$$\langle \mathbf{p}_1' \cdots \mathbf{p}_n'|T|\mathbf{p}_1 \cdots \mathbf{p}_n\rangle = \langle \mathbf{p}_1' \cdots \mathbf{p}_n'|R^\dagger T R|\mathbf{p}_1 \cdots \mathbf{p}_n\rangle \tag{2.9}$$
$$= \langle R\mathbf{p}_1' \cdots R\mathbf{p}_n'|T|R\mathbf{p}_1 \cdots R\mathbf{p}_n\rangle.$$

When the particles have nonzero spins the matrix elements of T are not invariant but covariant.

It is convenient to expand these matrix elements in a series expansion of scattering amplitudes between states with well-defined total angular momentum J, total spin S and orbital angular momentum ℓ. One of the reasons is because the two-body unitarity adopts a very simple form.

Let us consider a two-body state of particles with spins s_1 and s_2, that is characterized by the CM three-momentum $\mathbf{p}$ and the third components of spin σ_1 and σ_2 in their respective rest frames. This state is denoted by $|\mathbf{p}, \sigma_1\sigma_2\rangle$. Associated with this, we can define the two-body state with orbital angular momentum ℓ and third component of orbital angular momentum m, denoted by $|\ell m, \sigma_1\sigma_2\rangle$, as

$$|\ell m, \sigma_1\sigma_2\rangle = \frac{1}{\sqrt{4\pi}} \int d\hat{\mathbf{p}}\, Y_\ell^m(\hat{\mathbf{p}})|\mathbf{p}, \sigma_1\sigma_2\rangle. \tag{2.10}$$

Let us show first that this definition is meaningful because the state $|\ell m, \sigma_1\sigma_2\rangle$ transforms under the rotation group as the direct product of the irreducible representations associated with the orbital angular momentum ℓ and the spins s_1 and s_2 of the two particles.

We introduce a Lorentz transformation $U(\mathbf{p})$ along the vector $\mathbf{p}$ that takes the particle four-momentum at rest to its final value

$$U(\mathbf{p}) \begin{pmatrix} m \\ \mathbf{0} \end{pmatrix} = \begin{pmatrix} E_p \\ \mathbf{p} \end{pmatrix}, \tag{2.11}$$

with $E_p = \sqrt{m^2 + \mathbf{p}^2}$. We also introduce the rotation $R(\hat{\mathbf{p}})$ that takes $\hat{\mathbf{z}}$ to $\hat{\mathbf{p}}$,

$$R(\hat{\mathbf{p}})\hat{\mathbf{z}} = \hat{\mathbf{p}}. \tag{2.12}$$

In terms of the polar (θ) and azimuthal (ϕ) angles of $\hat{\mathbf{p}}$ this rotation is defined as

$$R(\hat{\mathbf{p}}) = R_z(\phi) R_y(\theta), \tag{2.13}$$

with the subscripts z and y indicating the axis of rotation. For latter convenience we write the Lorentz transformation $U(\mathbf{p})$ as

$$U(\mathbf{p}) = R(\hat{\mathbf{p}}) B_z(|\mathbf{p}|) R(\hat{\mathbf{p}})^{-1}, \tag{2.14}$$

where $B_z(|\mathbf{p}|)$ is a boost along the $\hat{\mathbf{z}}$ axis with velocity $v = -\beta$ and $\beta = |\mathbf{p}|/E_p$. Namely,

$$B_z(|\mathbf{p}|) = \begin{pmatrix} \gamma & 0 & 0 & \gamma\beta \\ 0 & 0 & 0 & 0 \\ 0 & 0 & 0 & 0 \\ \gamma\beta & 0 & 0 & \gamma \end{pmatrix} \tag{2.15}$$

and

$$\gamma = \frac{1}{\sqrt{1 - \beta^2}}. \tag{2.16}$$

Notice that one could also include any arbitrary rotation around the $\hat{\mathbf{z}}$ axis to the right end of Eq. (2.13). Of course, this does not have any affect on either Eqs. (2.12) and (2.14) (for the latter one let us note that $B_z(|\mathbf{p}|)$ commutes with a rotation around the $\hat{\mathbf{z}}$ axis).

Every monoparticle state $|\mathbf{p}, \sigma\rangle$ under the action of a rotation R transforms as

$$R|\mathbf{p}, \sigma\rangle = RU(\mathbf{p})|\mathbf{0}, \sigma\rangle = U(\mathbf{p}')U(\mathbf{p}')^{-1}RU(\mathbf{p})|\mathbf{0}, \sigma\rangle, \tag{2.17}$$

and $\mathbf{p}' = R\mathbf{p}$.[1] It is straightforward to show that

$$R = U(\mathbf{p}')^{-1}RU(\mathbf{p}). \tag{2.18}$$

---

[1] For a general Lorentz transformation these manipulations give rise to the Wigner rotation [5].

For that we explicitly write the Lorentz transformations $U(\mathbf{p}')$ and $U(\mathbf{p})$ as in Eq. (2.14) so that

$$U(\mathbf{p}')^{-1}RU(\mathbf{p}) = R(\hat{\mathbf{p}}')B_z(|\mathbf{p}|)^{-1}R(\hat{\mathbf{p}}')^{-1}RR(\hat{\mathbf{p}})B_z(|\mathbf{p}|)R(\hat{\mathbf{p}})^{-1}. \qquad (2.19)$$

Next, the product of rotations $R(\hat{\mathbf{p}}')^{-1}RR(\hat{\mathbf{p}})$ is a rotation around the z axis, $R_z(\gamma)$, since it leaves invariant $\hat{\mathbf{z}}$. Thus,

$$R(\hat{\mathbf{p}}')^{-1}RR(\hat{\mathbf{p}}) = R_z(\gamma), \qquad (2.20)$$

or, in other terms,

$$R(\hat{\mathbf{p}}') = RR(\hat{\mathbf{p}})R_z(\gamma)^{-1}, \qquad (2.21)$$

Taking into account Eqs. (2.20) and (2.21) in Eq. (2.19) it follows the result of Eq. (2.18) because $B_z(|\mathbf{p}|)$ and $R_z(\gamma)$ commute. Then, Eq. (2.17) implies that

$$R|\mathbf{p}, \sigma\rangle = U(\mathbf{p}')R|\mathbf{0}, \sigma\rangle = \sum_{\sigma'} D^{(s)}(R)_{\sigma'\sigma}|\mathbf{p}', \sigma'\rangle, \qquad (2.22)$$

with $D^{(s)}(R)$ the rotation matrix in the irreducible representation of the rotation group with spin s.

Now, it is clear from this result that the action of the rotation R on the state $|\mathbf{p}, \sigma_1\sigma_2\rangle$, which is the direct product of the states $|\mathbf{p}, \sigma_1\rangle$ and $|-\mathbf{p}, \sigma_2\rangle$, is

$$R|\mathbf{p}, \sigma_1\sigma_2\rangle = \sum_{\sigma_1', \sigma_2'} D^{(s_1)}(R)_{\sigma_1'\sigma_1} D^{(s_2)}(R)_{\sigma_2'\sigma_2}|\mathbf{p}', \sigma_1'\sigma_2'\rangle. \qquad (2.23)$$

We are now ready to derive the action of R on $|\ell m, \sigma_1\sigma_2\rangle$,

$$R|\ell m, \sigma_1\sigma_2\rangle = \sum_{\sigma_1', \sigma_2'} D^{(s_1)}(R)_{\sigma_1'\sigma_1} D^{(s_2)}(R)_{\sigma_2'\sigma_2} \frac{1}{\sqrt{4\pi}} \int d\hat{\mathbf{p}}' \, Y_\ell^m(R^{-1}\hat{\mathbf{p}}')|\mathbf{p}', \sigma_1'\sigma_2'\rangle$$

$$= \sum_{\sigma_1', \sigma_2', m'} D^{(\ell)}(R)_{m'm} D^{(s_1)}(R)_{\sigma_1'\sigma_1} D^{(s_2)}(R)_{\sigma_2'\sigma_2}|\ell m', \sigma_1'\sigma_2'\rangle. \qquad (2.24)$$

In this equation we have made use of the invariance of the solid-angle measure under rotations and the following property of the spherical harmonics:

$$Y_\ell^m(R^{-1}\hat{\mathbf{p}}') = \sum_{m'} D^{(\ell)}(R)_{m'm} Y_\ell^{m'}(\hat{\mathbf{p}}'). \qquad (2.25)$$

This property can be easily derived by noticing that

$$Y_\ell^m(\hat{\mathbf{p}}) = \langle \hat{\mathbf{p}} | \ell m \rangle, \tag{2.26}$$

where $|\mathbf{p}\rangle$ and $|\ell m\rangle$ are two generic states, the former having a well-defined orientation and the latter well-defined spin ℓ and third component of angular momentum m. Thus,

$$Y_\ell^m(R^{-1}\hat{\mathbf{p}}') = \langle R^{-1}\hat{\mathbf{p}}|\ell m\rangle = \langle \hat{\mathbf{p}}|R|\ell m\rangle = \sum_{m'} D_{m'm}^{(\ell)}(R) Y_\ell^{m'}(\hat{\mathbf{p}}). \tag{2.27}$$

Equation (2.24) shows that under a rotation the states defined in Eq. (2.10) has the right transformation under the action of a rotation R, and our proposition above [below Eq. (2.10)] is shown to hold.

Now, because of the transformation in Eq. (2.24), corresponding to the direct product of the spins s_1, s_2 and the orbital angular momentum ℓ, we can combine these angular momentum indices and end with the LSJ basis. In the latter every state is labeled by the total angular momentum J, the third component of the total angular momentum μ, the orbital angular momentum ℓ and the total spin S (resulting from the composition of the spins s_1 and s_2). Namely, we use the notation $|J\mu, \ell S\rangle$ for these states, which are then given by

$$|J\mu, \ell S\rangle = \sum_{\sigma_1, \sigma_2, m, M} (\sigma_1 \sigma_2 M | s_1 s_2 S)(m M \mu | \ell S J) | \ell m, \sigma_1 \sigma_2\rangle, \tag{2.28}$$

where we have introduced the standard Clebsch–Gordan coefficients for the composition of two angular momenta.[2] Next we introduce the isospin indices α_1 and α_2 corresponding to the third components of the isospins τ_1 and τ_2. This does not modify any of our previous considerations since isospin does not transform under the action of spatial rotations, although it is convenient for latter use. One can invert Eq. (2.10) and give the momentum-defined states in terms of those with well-defined orbital angular momentum,

$$|\mathbf{p}, \sigma_1 \sigma_2, \alpha_1 \alpha_2\rangle = \sqrt{4\pi} \sum_{\ell, m} Y_\ell^m(\hat{\mathbf{p}})^* |\ell m, \sigma_1 \sigma_2, \alpha_1 \alpha_2\rangle \tag{2.29}$$

$$= \sqrt{4\pi} \sum_{\substack{J, \mu, \ell, m \\ S, M, I, t_3}} Y_\ell^m(\hat{\mathbf{p}})^* (\sigma_1 \sigma_2 M | s_1 s_2 S)(m M \mu | \ell S J)(\alpha_1 \alpha_2 t_3 | \tau_1 \tau_2 I) | J\mu, \ell S, I t_3\rangle,$$

with I the total isospin of the particle pair and t_3 is the third component. The inversion of Eq. (2.29) provides us with the states $|J\mu, \ell S, I t_3\rangle$,

---

[2] The Clebsch–Gordan coefficient $(m_1 m_2 m_3 | j_1 j_2 j_3)$ is the composition for $\mathbf{j}_1 + \mathbf{j}_2 = \mathbf{j}_3$, with m_i referring to the third components of the spins.

$$|J\mu, \ell S, It_3\rangle = \frac{1}{\sqrt{4\pi}} \sum_{\substack{\sigma_1, \sigma_2 \\ M, m \\ \alpha_1, \alpha_2}} \int d\hat{\mathbf{p}} Y_\ell^m(\hat{\mathbf{p}})(\sigma_1\sigma_2 M|s_1s_2 S)(mM\mu|\ell SJ) \qquad (2.30)$$

$$\times (\alpha_1\alpha_2 t_3|\tau_1\tau_2 I)|\mathbf{p}, \sigma_1\sigma_2, \alpha_1\alpha_2\rangle.$$

From the normalization of the monoparticle states, Eq. (1.2), a two-body particle state with definite three-momentum satisfies the normalization

$$\langle \mathbf{p}', \sigma_1'\sigma_2', \alpha_1'\alpha_2'|\mathbf{p}, \sigma_1\sigma_2, \alpha_1\alpha_2\rangle = \frac{16\pi^2\sqrt{s}}{|\mathbf{p}|}\delta(\hat{\mathbf{p}}' - \hat{\mathbf{p}}), \qquad (2.31)$$

where we have dropped out a factor $(2\pi)^4\delta^{(4)}(p' - p)$. The total energy conservation guarantees that the moduli of the final and initial three-momenta in Eq. (2.31) are the same, that we denote by $|\mathbf{p}|$. In terms of this result and Eq. (2.30) it follows straightforwardly by taking into account the orthogonality properties of Clebsch–Gordan coefficients and spherical harmonics that

$$\langle J'\mu', \ell'S', I't_3'|J\mu, \ell S, It_3\rangle = \frac{4\pi\sqrt{s}}{|\mathbf{p}|}\delta_{J'J}\delta_{\mu'\mu}\delta_{\ell'\ell}\delta_{S'S}\delta_{I'I}\delta_{t_3't_3}. \qquad (2.32)$$

We are interested in the partial-wave amplitude (PWA) corresponding to the transition between states with quantum numbers $J\bar{\ell}\bar{S}I$ to states $J\ell SI$, that is given by the matrix element

$$T_{\ell S;\bar{\ell}\bar{S}}^{(JI)} = \langle J\mu, \ell S, It_3|T|J\mu, \bar{\ell}\bar{S}, It_3\rangle. \qquad (2.33)$$

Here we take the convention that the quantum numbers referring to the initial state are barred. Of course, the matrix element in Eq. (2.33) is independent of μ and t_3 because of the invariance under rotations in ordinary and isospin spaces, respectively. We can calculate this scattering matrix element in terms of those in the basis with definite three-momentum by replacing in Eq. (2.33) the states in the LSJ basis as given in Eq. (2.30). We then obtain in a first step

$$T_{\ell S;\bar{\ell}\bar{S}}^{(JI)} = \frac{1}{4\pi} \sum \int d\hat{\mathbf{p}} \int d\hat{\mathbf{p}}' \, Y_\ell^m(\hat{\mathbf{p}}')^* Y_{\bar{\ell}}^{\bar{m}}(\hat{\mathbf{p}})(\sigma_1\sigma_2 M|s_1s_2 S)(mM\mu|\ell SJ)(\alpha_1\alpha_2 t_3|\tau_1\tau_2 I)$$

$$\times (\bar{\sigma}_1\bar{\sigma}_2\bar{M}|\bar{s}_1\bar{s}_2 S)(\bar{m}\bar{M}\mu|\bar{\ell}\bar{S}J)(\bar{\alpha}_1\bar{\alpha}_2 t_3|\bar{\tau}_1\bar{\tau}_2 I)_s\langle \mathbf{p}', \sigma_1\sigma_2, \alpha_1\alpha_2|T|\mathbf{p}, \bar{\sigma}_1\bar{\sigma}_2, \bar{\alpha}_1\bar{\alpha}_2\rangle_s,$$
$$(2.34)$$

Here we have not shown the explicit indices over which the sum is done in order not to overload the notation.[3] We use next the rotation invariance of the T-matrix operator T to simplify the previous integral so that, at the end, we have just the integration over the final three-momentum solid angle. There are several steps involved that

---

[3]They correspond to those indicated under the summation symbol in Eq. (2.30) both for the initial and final states.

we discuss in detail. The referred rotational invariance of T implies that it remains invariant under the transformation $T \to R(\hat{\mathbf{p}}) T R(\hat{\mathbf{p}})^\dagger$, which implies at the level of the matrix elements that

$$\langle \mathbf{p}', \sigma_1\sigma_2, \alpha_1\alpha_2 | T | \mathbf{p}, \bar{\sigma}_1\bar{\sigma}_2, \bar{\alpha}_1\bar{\alpha}_2 \rangle = \langle \mathbf{p}', \sigma_1\sigma_2, \alpha_1\alpha_2 | R(\hat{\mathbf{p}}) T R(\hat{\mathbf{p}})^\dagger | \mathbf{p}, \bar{\sigma}_1\bar{\sigma}_2, \bar{\alpha}_1\bar{\alpha}_2 \rangle. \tag{2.35}$$

Under the action of the rotation $R(\hat{\mathbf{p}})^\dagger$ ($R(\hat{\mathbf{p}})^\dagger \hat{\mathbf{p}} = \hat{\mathbf{z}}$ and $R(\hat{\mathbf{p}})^\dagger \hat{\mathbf{p}}' = \hat{\mathbf{p}}''$) the final and initial states transform as, cf. Eq. (2.23),

$$R(\hat{\mathbf{p}})^\dagger | \mathbf{p}, \bar{\sigma}_1\bar{\sigma}_2, \bar{\alpha}_1\bar{\alpha}_2 \rangle = \sum_{\bar{\sigma}_1', \bar{\sigma}_2'} D^{(\bar{s}_1)}_{\bar{\sigma}_1'\bar{\sigma}_1}(R^\dagger) D^{(\bar{s}_2)}_{\bar{\sigma}_2'\bar{\sigma}_2}(R^\dagger) | \hat{\mathbf{z}}, \bar{\sigma}_1'\bar{\sigma}_2', \bar{\alpha}_1\bar{\alpha}_2 \rangle,$$

$$R(\hat{\mathbf{p}})^\dagger | \mathbf{p}', \sigma_1\sigma_2, \alpha_1\alpha_2 \rangle = \sum_{\sigma_1', \sigma_2'} D^{(s_1)}_{\sigma_1'\sigma_1}(R^\dagger) D^{(s_2)}_{\sigma_2'\sigma_2}(R^\dagger) | \hat{\mathbf{p}}'', \sigma_1'\sigma_2', \alpha_1\alpha_2 \rangle, \tag{2.36}$$

with the convention that R inside the argument of the rotation matrices refers to $R(\hat{\mathbf{p}})$. We insert Eqs. (2.35) and (2.36) into Eq. (2.34), and next transform $\hat{\mathbf{p}}' \to \hat{\mathbf{p}}''$ as integrations variables, take into account the invariance of the solid angle measure under such rotation and use Eq. (2.25) for

$$Y_{\bar{\ell}}^{\bar{m}}(\hat{\mathbf{p}}) = Y_{\bar{\ell}}^{\bar{m}}(R(\hat{\mathbf{p}})\hat{\mathbf{z}}) = \sum_{\bar{m}'} D^{(\bar{\ell})}_{\bar{m}'\bar{m}}(R^\dagger) Y_{\bar{\ell}}^{\bar{m}'}(\hat{\mathbf{z}}),$$

$$Y_{\ell}^{m}(\hat{\mathbf{p}}') = Y_{\ell}^{m}(R(\hat{\mathbf{p}})\hat{\mathbf{p}}'') = \sum_{m'} D^{(\ell)}_{m'm}(R^\dagger) Y_{\ell}^{m'}(\hat{\mathbf{p}}''). \tag{2.37}$$

Then, Eq. (2.34) for $T_{\ell S; \bar{\ell}\bar{S}}^{(JI)}$ can be rewritten as

$$T_{\ell S; \bar{\ell}\bar{S}}^{(JI)} = \frac{1}{4\pi} \sum \int d\hat{\mathbf{p}} \int d\hat{\mathbf{p}}'' (\sigma_1\sigma_2 M | s_1 s_2 S)(mM\mu | \ell S J)(\alpha_1\alpha_2 t_3 | \tau_1\tau_2 I) D^{(s_1)}_{\sigma_1'\sigma_1}(R^\dagger)^*$$

$$\times D^{(s_2)}_{\sigma_2'\sigma_2}(R^\dagger)^* D^{(\ell)}_{m'm}(R^\dagger)^* Y_{\ell}^{m'}(\hat{\mathbf{p}}'')^* (\bar{\sigma}_1\bar{\sigma}_2\bar{M} | \bar{s}_1\bar{s}_2\bar{S})(\bar{m}\bar{M}\mu | \bar{\ell}\bar{S}J)(\bar{\alpha}_1\bar{\alpha}_2 t_3 | \bar{\tau}_1\bar{\tau}_2 I)$$

$$\times D^{(\bar{s}_1)}_{\bar{\sigma}_1'\bar{\sigma}_1}(R^\dagger) D^{(\bar{s}_2)}_{\bar{\sigma}_2'\bar{\sigma}_2}(R^\dagger) D^{(\bar{\ell})}_{\bar{m}'\bar{m}}(R^\dagger) Y_{\bar{\ell}}^{\bar{m}'}(\hat{\mathbf{z}}) \langle \mathbf{p}'', \sigma_1'\sigma_2', \alpha_1\alpha_2 | \hat{T} | |\mathbf{p}|\hat{\mathbf{z}}, \bar{\sigma}_1'\bar{\sigma}_2', \bar{\alpha}_1\bar{\alpha}_2 \rangle. \tag{2.38}$$

From the composition of two rotation matrices it follows the result [5, 6]

$$\sum_{m_1, m_2} (m_1 m_2 M | \ell_1 \ell_2 L) D^{(\ell_1)}_{m_1'm_1}(R) D^{(\ell_2)}_{m_2'm_2}(R) = \sum_{M'} (m_1' m_2' M' | \ell_1 \ell_2 L) D^{(L)}_{M'M}(R). \tag{2.39}$$

We apply this rule first to the following two combinations in Eq. (2.38):

$$\sum_{\sigma_1,\sigma_2}(\sigma_1\sigma_2 M|s_1 s_2 S)D^{(s_1)}_{\sigma'_1\sigma_1}(R^\dagger)D^{(s_2)}_{\sigma'_2\sigma_2}(R^\dagger)=\sum_{M'}(\sigma'_1\sigma'_2 M'|s_1 s_2 S)D^{(S)}_{M'M}(R^\dagger)$$

$$\sum_{\bar\sigma_1,\bar\sigma_2}(\bar\sigma_1\bar\sigma_2\bar M|\bar s_1\bar s_2\bar S)D^{(\bar s_1)}_{\bar\sigma'_1\bar\sigma_1}(R^\dagger)D^{(\bar s_2)}_{\bar\sigma'_2\bar\sigma_2}(R^\dagger)=\sum_{\bar M'}(\bar\sigma'_1\bar\sigma'_2\bar M'|\bar s_1\bar s_2\bar S)D^{(\bar S)}_{\bar M'\bar M}(R^\dagger), \quad (2.40)$$

so that Eq. (2.38) becomes

$$T^{(JI)}_{\ell S;\bar\ell\bar S}=\frac{1}{4\pi}\sum\int d\hat{\mathbf{p}}\int d\hat{\mathbf{p}}''(\sigma'_1\sigma'_2 M'|s_1 s_2 S)D^{(S)}_{M'M}(R^\dagger)^*D^{(\ell)}_{m'm}(R^\dagger)^*(mM\mu|\ell SJ)$$

$$\times\,(\alpha_1\alpha_2 t_3|\tau_1\tau_2 I)Y^{m'}_\ell(\hat{\mathbf{p}}'')^*(\bar\sigma'_1\bar\sigma'_2\bar M'|\bar s_1\bar s_2\bar S)D^{(\bar S)}_{\bar M'\bar M}(R^\dagger)D^{(\bar\ell)}_{\bar m'\bar m}(R^\dagger)(\bar m\bar M\mu|\bar\ell\bar SJ)$$

$$\times\,(\bar\alpha_1\bar\alpha_2 t_3|\bar\tau_1\bar\tau_2 I)Y^{\bar m'}_{\bar\ell}(\hat{\mathbf{z}})\langle\mathbf{p}'',\sigma'_1\sigma'_2,\alpha_1\alpha_2|T||\mathbf{p}|\hat{\mathbf{z}},\bar\sigma'_1\bar\sigma'_2,\bar\alpha_1\bar\alpha_2\rangle. \quad (2.41)$$

The same relation in Eq. (2.39) is applied once more to the combinations in Eq. (2.41):

$$\sum_{m,M}(mM\mu|\ell SJ)D^{(S)}_{M'M}(R^\dagger)D^{(\ell)}_{m'm}(R^\dagger)=\sum_{\mu'}(m'M'\mu'|\ell SJ)D^{(J)}_{\mu'\mu}(R^\dagger),$$

$$\sum_{\bar m,\bar M}(\bar m\bar M\mu|\bar\ell\bar SJ)D^{(\bar S)}_{\bar M'\bar M}(R^\dagger)D^{(\bar\ell)}_{\bar m'\bar m}(R^\dagger)=\sum_{\bar\mu'}(\bar m'\bar M'\bar\mu'|\bar\ell\bar SJ)D^{(J)}_{\bar\mu'\mu}(R^\dagger). \quad (2.42)$$

We take Eq. (2.42) into Eq. (2.41) which now reads

$$T^{(JI)}_{\ell S;\bar\ell\bar S}=\frac{1}{4\pi}\sum\int d\hat{\mathbf{p}}\int d\hat{\mathbf{p}}''(\sigma'_1\sigma'_2 M'|s_1 s_2 S)(m'M'\mu'|\ell SJ)(\alpha_1\alpha_2 t_3|\tau_1\tau_2 I)Y^{m'}_\ell(\hat{\mathbf{p}}'')^*$$

$$\times\,(\bar\sigma'_1\bar\sigma'_2\bar M'|\bar s_1\bar s_2\bar S)(\bar m'\bar M'\bar\mu'|\bar\ell\bar SJ)(\bar\alpha_1\bar\alpha_2 t_3|\bar\tau_1\bar\tau_2 I)Y^{\bar m'}_{\bar\ell}(\hat{\mathbf{z}})D^{(J)}_{\mu'\mu}(R^\dagger)^*D^{(J)}_{\bar\mu'\mu}(R^\dagger)$$

$$\times\,\langle\mathbf{p}'',\sigma'_1\sigma'_2,\alpha_1\alpha_2|T||\mathbf{p}|\hat{\mathbf{z}},\bar\sigma'_1\bar\sigma'_2,\bar\alpha_1\bar\alpha_2\rangle. \quad (2.43)$$

Now, the partial-wave amplitude $T^{(JI)}_{\ell S;\bar\ell\bar S}$ is independent of μ and then

$$T^{(JI)}_{\ell S;\bar\ell\bar S}=\frac{1}{2J+1}\sum_{\mu=-J}^{J}T^{(JI)}_{\ell S;\bar\ell\bar S}. \quad (2.44)$$

Thus, the same result in Eq. (2.43) is obtained with the product $D^{(J)}_{\mu'\mu}(R^\dagger)^*D^{(J)}_{\bar\mu'\mu}(R^\dagger)$ replaced by

$$\frac{1}{2J+1}\sum_{\mu=-J}^{J}D^{(J)}_{\mu'\mu}(R^\dagger)^*D^{(J)}_{\bar\mu'\mu}(R^\dagger)=\frac{\delta_{\bar\mu'\mu'}}{2J+1}, \quad (2.45)$$

as follows from the unitarity character of the rotation matrices. As a consequence any dependence in $\hat{\mathbf{p}}$ present in the integrand of Eq. (2.43) disappears in the average of Eq. (2.44), the integration in the solid angle $\hat{\mathbf{p}}$ is trivial and gives a factor 4π.

Taking into account the Kronecker delta in Eq. (2.45) for the third component of the total angular momentum, and a new one that arises because $Y_{\bar{\ell}}^{\bar{m}'}(\hat{\mathbf{z}})$ is not zero only for $\bar{m}' = 0$, we end with the following expression for $T_{\ell S; \bar{\ell} \bar{S}}^{(JI)}$:

$$T_{\ell S; \bar{\ell} \bar{S}}^{(JI)} = \frac{Y_{\bar{\ell}}^0(\hat{\mathbf{z}})}{2J+1} \sum_{\substack{\sigma_1, \sigma_2, \bar{\sigma}_1 \\ \bar{\sigma}_2, \alpha_1, \alpha_2 \\ \bar{\alpha}_1, \bar{\alpha}_2, m}} \int d\hat{\mathbf{p}}'' \langle \mathbf{p}'', \sigma_1\sigma_2, \alpha_1\alpha_2 | T | \mathbf{p}|\hat{\mathbf{z}}, \bar{\sigma}_1\bar{\sigma}_2, \bar{\alpha}_1\bar{\alpha}_2\rangle \quad (2.46)$$

$$\times Y_{\ell}^m(\hat{\mathbf{p}}'')^*(\sigma_1\sigma_2 M|s_1s_2 S)(mM\bar{M}|\ell SJ)(\bar{\sigma}_1\bar{\sigma}_2\bar{M}|\bar{s}_1\bar{s}_2\bar{S})(0\bar{M}\bar{M}|\bar{\ell}\bar{S}J)$$

$$\times (\alpha_1\alpha_2 t_3|\tau_1\tau_2 I)(\bar{\alpha}_1\bar{\alpha}_2 t_3|\bar{\tau}_1\bar{\tau}_2 I),$$

where we have removed the primes on top of the spin and orbital angular momentum third component symbols and in the previous sum $M = \sigma_1 + \sigma_2$ and $\bar{M} = \bar{\sigma}_1 + \bar{\sigma}_2$.

Next, we derive the unitarity relation corresponding to our normalization for the partial-wave projected amplitudes $T_{\ell S; \bar{\ell} \bar{S}}^{(JI)}$. For that we employ the unitarity relation for the T matrix of Eq. (1.8). By time-reversal invariance it follows that the partial-wave amplitudes are symmetric (this is demonstrated in the footnote 9 of [7], see also chapters 3 and 5 of Ref. [5]) and therefore

$$2\mathrm{Im}\, T_{\ell S; \bar{\ell} \bar{S}}^{(JI)} = \langle J\mu, \ell S, Tt_3 | TT^\dagger | J\mu, \bar{\ell}\bar{S}, Tt_3\rangle. \quad (2.47)$$

On the rhs we introduce between T and $T^\dagger$ a two-body resolution of the identity of states $|J\mu, \ell S, It_3\rangle$ (we have restricted our vector space to the one generated by these states, below the threshold of multiparticle production) such that, taking into account their normalization in Eq. (2.32), one ends with

$$\mathrm{Im}\, T_{\ell S; \bar{\ell} \bar{S}}^{(JI)} = \sum_{\ell'', S''} \frac{|\mathbf{p}''|}{8\pi\sqrt{s}} T_{\ell, S; \ell'', S''}^{(JI)} T_{\ell'', S''; \bar{\ell}\bar{S}}^{(JI)*}. \quad (2.48)$$

The phase-space factor is included in the diagonal matrix

$$\rho_{ij} = \frac{|\mathbf{p}|_i}{8\pi\sqrt{s}}\delta_{ij}. \quad (2.49)$$

Another form of expressing the unitarity requirement of Eq. (2.48) is by introducing the inverse of the matrix $T^{(JI)}(s)$. Employing a matrix notation we rewrite this equation as

$$T^{(JI)} - T^{\dagger(JI)} = 2i T^{(JI)}\rho T^{\dagger(JI)}. \quad (2.50)$$

Multiplying to the left by $T^{(JI)^{-1}}$ and to the right by $T^{\dagger(JI)^{-1}}$, we arrive to the desired result

$$\Im T^{(JI)^{-1}}(s) = -\rho(s), \tag{2.51}$$

above the thresholds of the channels involved. Along this work, we also denote by channel any of the states interacting in a given process.

A more standard definition of the S-matrix for partial waves implies to redefine it as

$$S^{(JI)} = I + 2i\rho^{\frac{1}{2}}T^{(JI)}\rho^{\frac{1}{2}}. \tag{2.52}$$

This redefinition amounts to evaluate the S matrix between partial-wave projected states

$$\sqrt{\frac{|\mathbf{p}|_i}{4\pi}}|J\mu, \ell S; i\rangle, \tag{2.53}$$

which are just normalized to the product of Kronecker deltas with unit coefficient, instead of the original normalization in Eq. (2.32). As a result now the diagonal matrix elements of the identity operator I and S_J are just 1 and $\eta_i e^{2i\delta_i}$. Namely,

$$S_{ii}^{(JI)} = \eta_i e^{2i\delta_i}, \tag{2.54}$$

where η_i is the inelasticity for channel i and δ_i its phase shift. We have included the superscripts I and J to emphasize that we are considering a certain set of partial waves with definite total angular momentum J and total isospin I.

A slight change in the formalism for projecting into partial-wave amplitudes is needed when the two-body states comprise two identical particles or these two particles can be treated as identical ones within the isospin formalism. This has been treated in detail for fermions and bosons in Refs. [8, 9], in order. These (anti)symmetric states are defined by

$$|\mathbf{p}, \sigma_1\sigma_2, \alpha_1\alpha_2\rangle_S = \frac{1}{\sqrt{2}}\left(|\mathbf{p}, \sigma_1\sigma_2, \alpha_1\alpha_2\rangle \pm |-\mathbf{p}, \sigma_2\sigma_1, \alpha_2\alpha_1\rangle\right), \tag{2.55}$$

with the subscript S indicating the (anti)symmetrized nature of the state under the exchange of the two particles, the $+$ is for bosons and the $-$ is for fermions. Making use of Eq. (2.29) we can write the (anti)symmetric states in terms of those with well-defined J. As a result, instead of Eq. (2.30) for isolating the states with well-defined J, we have now

$$|\mathbf{p}, \sigma_1\sigma_2, \alpha_1\alpha_2\rangle_S = \sqrt{4\pi} \sum_{\substack{J, \mu, \ell, m \\ S, M, I, t_3}} \frac{1 \pm (-1)^{\ell+S+I}}{\sqrt{2}} (\sigma_1\sigma_2 M|s_1s_2 S)(mM\mu|\ell S J)$$

$$\tag{2.56}$$

$$\times (\alpha_1\alpha_2 t_3|\tau_1\tau_2 I) Y_\ell^m(\hat{\mathbf{p}})^* |J\mu, \ell S, I t_3\rangle.$$

In deducing this expression we have taken into account the following symmetric properties of the Clebsch–Gordan coefficients:

$$(\sigma_2\sigma_1 M|s_2s_1 S) = (-1)^{S-s_1-s_2}(\sigma_1\sigma_2 M|s_1s_2 S),$$
$$(\alpha_2\alpha_1 t_3|t_2 t_1 I) = (-1)^{I-t_1-t_2}(\alpha_1\alpha_2 t_3|\tau_1\tau_2 I),$$
$$Y_\ell^m(-\hat{\mathbf{p}}) = (-1)^\ell Y_\ell^m(\hat{\mathbf{p}}).$$

$$\tag{2.57}$$

Of course, due to the fact that we are dealing with indistinguishable particles within the isospin formalism it follows that $s_1 = s_2$, $\tau_1 = \tau_2$. The combination $(1 \pm (-1)^{\ell+S+I})/\sqrt{2}$ in Eq. (2.56) is denoted in the following as $\chi(\ell S I)$ and takes into account the (anti)symmetric character of the two particles, so that only states with (odd)even $\ell + S + I$ are allowed. The inversion of Eq. (2.56) gives (we assume in the following that $\ell + S + I$=even(odd) for bosons(fermions), so that $\chi(\ell S I) = \sqrt{2}$),

$$|J\mu, \ell S, I t_3\rangle = \frac{1}{\sqrt{8\pi}} \sum_{\substack{\sigma_1, \sigma_2 \\ M, m \\ \alpha_1, \alpha_2}} \int d\hat{\mathbf{p}} Y_\ell^m(\hat{\mathbf{p}})(\sigma_1\sigma_2 M|s_1s_2 S)(mM\mu|\ell S J)(\alpha_1\alpha_2 t_3|\tau_1\tau_2 I)$$

$$\times |\mathbf{p}, \sigma_1\sigma_2, \alpha_1\alpha_2\rangle_S.$$

$$\tag{2.58}$$

By comparing with Eq. (2.30) we conclude that the only difference of writing the states $|J\mu, \ell S, I t_3\rangle$ in terms of the (anti)symmetric ones is just a factor $1/\sqrt{2}$. Therefore, we can use the same expression of Eq. (2.46) for calculating the partial-wave amplitudes in terms of the (anti)symmetric states but including an extra factor $1/\sqrt{2}$ for every state obeying (anti)symmetric properties under the exchange of its two particles. This is the so-called unitary normalization introduced in Ref. [10].

Let us show how the unitarity relation stems within the less abstract and more definite setup of potential scattering, for more details the reader can consult section 2 of Ref. [7]. The full Hamiltonian H is the free one H_0 plus the potential v, $H = H_0 - v$. A minus sign in front of v is introduced to conform with the sign convention employed in the definition of the T matrix in Eq. (1.6). We denote by $r_0(z)$ and $r(z)$ the resolvents of H_0 and H, in order,

$$r_0(z) = (H_0 - z)^{-1},$$
$$r(z) = (H - z)^{-1},$$

$$\tag{2.59}$$

with $\Im z \neq 0$. The following equations for $r(z)$ follow,

$$r(z) = r_0(z) + r_0(z)vr(z) \tag{2.60}$$

$$= r_0(z) + r(z)vr_0(z). \tag{2.61}$$

For instance, take the following steps $r(z) = r_0(I - vr_0)^{-1} = r_0(I - vr_0 + vr_0)(I - vr_0)^{-1} = r_0 + r_0vr(z)$. For the other equation multiply the previous derivation by $I - vr_0$ to the right.

The relation between the T matrix $T(z)$ and the resolvent $r(z)$ is, by definition,

$$T(z)r_0(z) = vr(z), \tag{2.62}$$

such that from Eq. (2.60)

$$r(z) = r_0(z) + r_0(z)T(z)r_0(z). \tag{2.63}$$

By comparing this result with Eq. (2.61) we also obtain that

$$r_0(z)T(z) = r(z)v. \tag{2.64}$$

The Lippmann–Schwinger (LS) equation results by employing Eq. (2.60) for $r(z)$, and its relation with $T(z)$, Eq. (2.62), such that

$$T(z) = v + vr_0(z)T(z). \tag{2.65}$$

Had we used instead Eqs. (2.61) and (2.64) we would have obtained this other form of the LS equation

$$T(z) = v + T(z)r_0(z)v. \tag{2.66}$$

An interesting property of $T(z)$ is that

$$T(z)^\dagger = T(z^*) \tag{2.67}$$

as follows from the fact that $r(z)^\dagger = r(z^*)$, as it is clear from its definition.[4] The previous equation illustrates as well the Hermitian unitarity [2].

We can obtain too a relationship between two resolvent operators evaluated at different values of z, which is known as the Hilbert identity, and that it is very useful also to establish the unitarity properties of the T matrix. The Hilbert identity reads

$$r(z_1) - r(z_2) = (z_1 - z_2)r(z_1)r(z_2). \tag{2.68}$$

---

[4]The potential is also required to fulfill this property, $v(z)^\dagger = v(z^*)$.

For its demonstration take the difference $r(z_2)^{-1} - r(z_1)^{-1} = z_1 - z_2$ and multiply it to the left by $r(z_1)$ and to the right by $r(z_2)$. In terms of the T matrix the Hilbert identity reads

$$T(z_1) - T(z_2) = (z_1 - z_2)T(z_1)r_0(z_1)r_0(z_2)T(z_2), \qquad (2.69)$$

as follows by multiplying Eq. (2.68) to the left and right by v. For the left-hand side (lhs) in Eq. (2.68) use the LS equation (2.65), because then $vr(z_1)v - vr(z_2)v = vr_0(z_1)T(z_1) - vr_0(z_2)T(z_2) = T(z_1) - T(z_2)$. For the rhs just employ the Eqs. (2.62) and (2.64), $vr(z_1)r(z_2)v = T(z_1)r_0(z_1)r_0(z_2)T(z_2)$.

The partial-wave projected LS equation arises by considering two-body states projected in a given partial wave, $|k, \ell S, J\mu\rangle$, where k refers to the modulus of the three-momentum, as introduced in Eq. (2.31). To shorten the notation, we typically denote the discrete indices globally as λ, and then write $|k, \lambda\rangle$ for the same state. This compact notation is also valuable because it readily shows that the results can be applied to other choice of partial-wave projection, e.g., in the helicity basis. The partial-wave states are normalized such that

$$\langle k, \lambda | k', \lambda' \rangle = 2\pi^2 \frac{\delta(k - k')}{k^2} \delta_{\lambda\lambda'}. \qquad (2.70)$$

This normalization is consistent with that of Eq. (2.32) if $\sqrt{s} = p_1^0 + p_2^0$ is replaced by $m_1 + m_2$. Furthermore, we also have to multiply this equation by $2\pi\delta(k^2/2\mu - k'^2/2\mu) = 2\pi\mu\delta(k - k')/k$, because the LS equation is three-dimensional. Here $\mu = m_1 m_2/(m_1 + m_2)$ is the reduced mass. Finally, we divide the result by $4m_1 m_2$, because in a nonrelativistic theory a plane wave is normalized as $(2\pi)^3\delta(\mathbf{k} - \mathbf{k}')$, without the factor $2p_1^0$ (which in the nonrelativistic limit becomes simply $2m_1$).

The matrix elements of the T matrix between partial-wave states constitute the partial-wave amplitudes, indicated schematically by $T_{ij}(k, k'; z)$, and corresponding to

$$T_{ij}(k, k'; z) = \langle k, \lambda_i | t(z) | k', \lambda_j \rangle. \qquad (2.71)$$

The LS equation in partial waves is obtained by taking the matrix element between partial-wave states of Eqs. (2.65) and (2.66),

$$\begin{aligned}
T_{ij}(k, k'; z) &= v_{ij}(k, k') + \frac{\mu}{\pi^2} \sum_n \int_0^\infty \frac{dq\, q^2}{q^2 - 2\mu z} v_{in}(k, q) T_{nj}(q, k'; z) \\
&= v_{ij}(k, k') + \frac{\mu}{\pi^2} \sum_n \int_0^\infty \frac{dq\, q^2}{q^2 - 2\mu z} T_{in}(k, q; z) v_{nj}(q, k'). \quad (2.72)
\end{aligned}$$

In this equation we have included between the operators $T(z)$ and v an intermediate set of two-body partial-wave states $|q, \lambda_j\rangle$ taking into account their normalization in Eq. (2.70), and the notation

$$v_{ij}(k, k') = \langle k, \lambda_i | v | k', \lambda_j \rangle. \tag{2.73}$$

For the Hilbert identity in partial waves we have from Eq. (2.69) that

$$T_{ij}(k, k'; z_1) - T_{ij}(k, k'; z_2) = (z_1 - z_2) \frac{2\mu^2}{\pi^2} \sum_n \int_0^\infty \frac{dq\, q^2}{(q^2 - 2\mu z_1)(q^2 - 2\mu z_2)}$$
$$\times\, T_{in}(k, q; z_1) T_{nj}(q, k'; z_2). \tag{2.74}$$

Next, we use the property that follows from Eq. (2.67),

$$T_{ij}(k, k'; z^*) = T_{ji}(k', k; z)^*, \tag{2.75}$$

with $z_2 = z_1^* = z^*$ in Eq. (2.74). It results

$$T_{ij}(k, k'; z) - T_{ji}(k', k; z)^* = 2i\Im z \frac{2\mu^2}{\pi^2} \sum_n \int_0^\infty \frac{dq\, q^2}{(q^2 - 2\mu z)(q^2 - 2\mu z^*)}$$
$$\times\, T_{in}(k, q; z) T_{nj}(q, k'; z)^*. \tag{2.76}$$

Assuming time-reversal invariance the PWAs are symmetric, $T_{ij}(k, k'; z) = T_{ji}(k', k; z)$, and then Eq. (2.76) simplifies as

$$\Im T_{ij}(k, k'; z) = \Im z \frac{2\mu^2}{\pi^2} \sum_n \int_0^\infty \frac{dq\, q^2}{(q^2 - 2\mu z)(q^2 - 2\mu z^*)} T_{in}(k, q; z) T_{jn}(k', q; z)^*$$
$$= \Im z \frac{2\mu^2}{\pi^2} \sum_n \int_0^\infty \frac{dq\, q^2}{(q^2 - 2\mu\Re z)^2 + (2\mu\Im z)^2} T_{in}(k, q; z) T_{jn}(k', q; z)^* \tag{2.77}$$

We now take the limit $\Im z \to 0^+$ and use the result

$$\lim_{\Im z \to 0^+} \frac{2\mu\Im z}{(q^2 - 2\mu\Re z)^2 + (2\mu\Im z)^2} = \pi\delta(q^2 - 2\mu\Re z). \tag{2.78}$$

We indicate by κ the on-shell momentum,

$$\kappa = \sqrt{2\mu\Re z}, \tag{2.79}$$

and then Eq. (2.77) reads, with $z = E + i\varepsilon$ and $\varepsilon \to 0^+$,

$$\Im T_{ij}(k, k'; z) = \theta(E) \frac{\mu\kappa}{2\pi} \sum_n T_{in}(k, \kappa; z) T_{jn}(k', \kappa; z)^*, \tag{2.80}$$

which is the off-shell unitarity relation.

Two important particular cases of Eq. (2.80) are the half-off-shell and on-shell unitarity relations. For the former, one takes $E = k'^2/2\mu$ (so that $\kappa = k'$) in Eq. (2.80), which then reads

$$\Im T_{ij}(k, \kappa; E + i\varepsilon) = \theta(E)\frac{\mu\kappa}{2\pi}\sum_n T_{in}(k, \kappa; E + i\varepsilon)T_{jn}(\kappa, \kappa; E + i\varepsilon)^*. \quad (2.81)$$

This is a unitarity relation of the same type as the one for form factors [5, 11], that are discussed in Chap. 13. In connection with this, we have here another version of the Watson final-state theorem because Eq. (2.81) implies that along the RHC the phase of the half-off-shell PWA is the same (modulo π) as the phase of the on-shell PWA for the uncoupled case, since the lhs of Eq. (2.81) is real.

The on-shell unitarity relation, or simply unitarity, stems by taking the extra requirement $k' = k$ in the half-off-shell case. Then Eq. (2.81) becomes

$$\Im T_{ij}(\kappa, \kappa; E + i\varepsilon) = \theta(E)\frac{\mu\kappa}{2\pi}\sum_n T_{in}(\kappa, \kappa; E + i\varepsilon)T_{jn}(\kappa, \kappa; E + i\varepsilon)^*. \quad (2.82)$$

This imaginary part is the reason of the presence of the right-hand cut (RHC) or unitarity cut in the PWAs for positive real values of the energy (or physical ones). This is clear if we take into account the Hermitian unitarity relation in Eq. (2.67) and the symmetric character of the partial-wave amplitudes, so that $T_{ij}(\kappa, \kappa, E + i\varepsilon) - T_{ij}(\kappa, \kappa, E - i\varepsilon) = 2i\Im T_{ij}(\kappa, \kappa, E + i\varepsilon)$, with the latter given in Eq. (2.82).

The partial-wave decomposition of the S matrix, given by its matrix elements between states $|k, \lambda\rangle$, is

$$S_{ij}(E) = \delta_{ij} + i\frac{\mu\kappa}{\pi}T_{ij}(\kappa, \kappa; E + i\varepsilon). \quad (2.83)$$

This is a unitary operator for $E \geq 0$ because it satisfies

$$S(E)S(E)^\dagger = S(E)^\dagger S(E) = I, \ E \geq 0, \quad (2.84)$$

as consequence of the on-shell unitarity in partial waves, Eq. (2.82).

Chapter 3
Crossing. Crossed-Channel Singularities

From perturbative QFT it is clear that a generic quantum filed $\phi_i(x)$ contains both the annihilation operators of a type of particles and the creation operators of the corresponding antiparticles [4]. The former term is multiplied by the space-time factor $\exp(-ipx)$ while the latter is so by $\exp(ipx)$. To get the basic idea involved in crossing let us consider that the field has spin zero. Therefore, the same vertices in a given scattering process can be associated with a particle of four-momentum p or with an antiparticle with four-momentum $-p$ and viceversa. This implies that if we have a scattering amplitude of the form

$$a_1(p_1) + a_2(p_2) + \cdots \rightarrow b_1(p_1') + b_2(p_2') + \cdots \tag{3.1}$$

the same scattering amplitude governs any other process in which one or several particles are changed from initial/final to final/initial and at the same time there is a flip in the global sign of the four-momenta. For instance, for the previous reaction we could have many others related by crossing, in particular

$$a_1(p_1) + a_2(p_2) + \cdots + \bar{b}(-p_1') \rightarrow b_2(p_2') + \cdots \tag{3.2}$$

where the bar indicates the corresponding antiparticle. This is the basic content of crossing.

Let us particularize crossing to the two-body scattering $a + b \rightarrow c + d$. We can then distinguish the following related processes:

$$a(p_1) + b(p_2) \rightarrow c(p_3) + d(p_4), \tag{3.3}$$
$$a(p_1) + \bar{c}(-p_3) \rightarrow \bar{b}(-p_2) + d(p_4), \tag{3.4}$$
$$a(p_1) + \bar{d}(-p_4) \rightarrow c(p_3) + \bar{b}(-p_2). \tag{3.5}$$

From top to bottom, these processes are referred to as s-channel, t-channel and u-channel, in order. We also denote the s-channel as the direct one while the t- and u-channels are also called crossed channels. Apart from the processes indicated

© The Author(s), under exclusive licence to Springer Nature Switzerland AG 2019   23
J. A. Oller, *A Brief Introduction to Dispersion Relations*, SpringerBriefs in Physics,
https://doi.org/10.1007/978-3-030-13582-9_3

in Eqs. (3.3)–(3.5), there are other three processes in which instead of exchanging $b(p_2) \rightarrow \bar{b}(-p2)$ from the initial to the final state, we could also exchange $a(p_1) \rightarrow \bar{a}(-p_1)$. These processes can also be obtained by CPT invariance from the ones shown in these equations.

Under the exchange of signs in the four-momenta, the s, t and u variables for every channel are related. Let us designate with a subscript t and u the Mandelstam variables for the t- and u-channels, respectively. Then we have for the t-channel:

$$s_t = (p_1 - p_3)^2 = t, \tag{3.6}$$

$$t_t = (p_1 + p_2)^2 = s,$$

$$u_t = (p_1 - p_4)^2 = u, \tag{3.7}$$

and for the u-channel the relations are

$$s_u = (p_1 - p_4)^2 = u, \tag{3.8}$$

$$t_u = (p_1 - p_3)^2 = t,$$

$$u_u = (p_1 + p_2)^2 = s. \tag{3.9}$$

The physical regions for these processes are disjoint. To simplify the discussion let us take that the four particles have the same mass m, e.g., this is the case of $\pi\pi$ scattering. The s, t and u variables are given in the CM by

$$s = 4(m^2 + \mathbf{p}^2), \tag{3.10}$$

$$t = -2(s/4 - m^2)(1 - \cos\theta),$$

$$u = -2(s/4 - m^2)(1 + \cos\theta),$$

with θ the scattering angle. From here we see that the physical region for the s-channel comprises the domain

$$s \geq 4m^2, \tag{3.11}$$

$$t \leq 0,$$

$$u \leq 0.$$

For the other channels the same values take place in terms of the variables with the subscripts. Thus, for the t-channel

$$t = s_t \geq 4m^2, \tag{3.12}$$

$$s = t_t \leq 0,$$

$$u = u_t \leq 0.$$

In turn for the u-channel,

$$u = s_u \geq 4m^2, \tag{3.13}$$
$$s = u_u \leq 0,$$
$$t = t_u \leq 0.$$

For the equal-mass case we have the relation, cf. Eq. (2.4),

$$s + t + u = 4m^2. \tag{3.14}$$

Therefore, only two of the three variables are independent.

Analyticity assumes that the scattering amplitudes in the three disjoint physical regions for the s-, t- and u-channels are given by the same analytical function $A(s, t, u)$ of s and t [u is then given by Eq. (3.14)]. The physical values for the different channels correspond to boundary values of this analytic function.

In particular, if we take a constant value of t the unitarity cut associated with the normal cuts in the u-channel, cf. Eq. (2.5), gives rise to a new cut in the complex s plane apart from the s-channel unitary cut. This is a simple example of a crossed-channel cut (also called unphysical cut, because it involves unphysical values of the Mandelstam variables in the s-channel) arising from a branch point singularity attached to a two-body threshold. In particular, this cut runs for $u \geq 4m^2$ so that it correspond to the s values

$$s = 4m^2 - t - u \leq -t, \tag{3.15}$$

and for $s \geq 4m^2$ we have the s-channel unitary cut.

For particles with spin the analytical continuation of the scattering amplitude in the complex s and t planes is more involved due to the presence of kinematical singularities, whose origin is not dynamical, like the unitarity cuts in the s or crossed channels. They have to do with the solutions of the relativistic equations for the particles with spins, like the spinors for spin 1/2. For a general account on kinematical singularities we refer to [12, 13].

A possible way to deal with the kinematics singularities is to isolate Lorentz invariant functions out of the scattering amplitudes. For instance, let us consider the process $\pi^a(q)N(p, \sigma; \alpha) \rightarrow \pi^{a'}(q')N(p', \sigma'; \alpha')$, where a and a' denote the Cartesian coordinates in the isospin space. In terms of them the charged pions correspond to the combination

$$\sum_a \frac{\pi^a \tau^a}{\sqrt{2}} = \begin{pmatrix} \frac{\pi^0}{\sqrt{2}} & \pi^+ \\ \pi^- & -\frac{\pi^0}{\sqrt{2}} \end{pmatrix}. \tag{3.16}$$

First, the scattering amplitude contains two invariant isospin amplitudes corresponding to $I = 1/2$ and 3/2, because the pions are isospin 1 particles and the nucleons have isospin 1/2. Any matrix in the isospin 1/2 space of nucleons can be expressed as a superposition of Pauli matrices τ^a and the 2×2 identity matrix. Given two pions with indices a and a' the tensors with good properties under isospin rotations at our disposition are $\delta_{aa'}$ and $[\tau_a, \tau_{a'}]$.[1] In this way we write,

$$T_{aa'} = \delta_{a'a}T^+ + \frac{1}{2}[\tau_a, \tau_{a'}]T^-. \qquad (3.17)$$

In turn the two amplitudes $T^\pm$ are operators acting in the space of the Dirac spinors and can be written as a linear combination of the 16 linearly independent matrices $I, \gamma^\mu, \sigma^{\mu\nu}, \gamma_5$ and $\gamma_5\gamma^\mu$ with the Lorentz indices contracted with four-momenta. The matrices γ_5 and $\gamma_5\gamma^\mu$ would violate parity and they do not appear. The set $\sigma^{\mu\nu}$ does not appear either because of the Gordon identity,[2] which implies that it does not give rise to any independent structure apart from the ones already accounted for by the identity matrix and γ^μ. Taking also into account that $\not{p}u(\mathbf{p}, \sigma) = mu(\mathbf{p}, \sigma)$ and $\bar{u}(\mathbf{p}', \sigma')\not{p}' = mu(\mathbf{p}', \sigma')$, we arrive to the standard form [12]

$$T^\pm = \bar{u}(p', \sigma')\left[A^\pm(s, t, u) + \frac{1}{2}(\not{q} + \not{q}')B^\pm(s, t, u)\right]u(p, \sigma). \qquad (3.19)$$

The analytical properties of the Lorentz invariant functions $A^\pm$ and $B^\pm$ are essentially the same as those of the scattering amplitude for scalar particles. The other factors in Eq. (3.19) have to be taken into account in establishing relations between analyticity and experimental results.

The crossed-channel poles, corresponding to poles in the crossed t- and u-channels for certain real values of t or u, in order, give rise to crossed cuts in the complex s plane of partial-wave amplitudes. For instance, consider the u-channel proton pole in $\pi^- n \to \pi^- n$ scattering (in this sense these poles are called "bound states", even though they could be elementary states or composite of other degrees of freedom [14]). Such a pole gives rise to a crossed cut in a given partial wave. For pion–nucleon scattering in the CM the u variable is given by (m and m_π are the nucleon and pion masses)

$$u = m^2 + m_\pi^2 - 2\omega E - 2\mathbf{p}^2\cos\theta. \qquad (3.20)$$

In this equation E and ω are the nucleon and pion CM energies, respectively. When performing the partial-wave projection the scattering angle is integrated and $\cos\theta \in [-1, 1]$. Thus, setting $u = m^2$ in Eq. (3.20) and expressing ω, E and $\mathbf{p}^2$ in terms of s,

---

[1] No tensor of rank should 2 be considered because its combination with an isospin 1/2 cannot get rise to an isospin 1/2.

[2] The Gordon identity establishes that

$$\bar{u}(\mathbf{p}', \sigma')\gamma^\mu u(\mathbf{p}, \sigma) = \frac{1}{2m}\bar{u}(\mathbf{p}', \sigma')\left[(p' + p)^\mu + i\sigma^{\mu\nu}(p' - p)_\nu\right]u(\mathbf{p}, \sigma). \qquad (3.18)$$

$$\omega = \frac{s + m_\pi^2 - m^2}{2\sqrt{s}}, \tag{3.21}$$

$$E = \frac{s + m^2 - m_\pi^2}{2\sqrt{s}},$$

$$\mathbf{p}^2 = \frac{\lambda(s, m^2, m_\pi^2)}{4s},$$

$$\lambda(s, m_1^2, m_2^2) = (s - (m_1 + m_2)^2)(s - (m_1 - m_2)^2),$$

with $\lambda(s, m_1^2, m_2^2)$ the Källén triangle function, we have the following solutions for s as a function of $x = \cos\theta$,

$$s_1(x) = \frac{m^2 x + m_\pi^2(1 + x) - \sqrt{m^4 + 2m_\pi^4(1 + x) + 2m^2 m_\pi^2(-1 + x + 2x^2)}}{1 + x}, \tag{3.22}$$

$$s_2(x) = \frac{m^2 x + m_\pi^2(1 + x) + \sqrt{m^4 + 2m_\pi^4(1 + x) + 2m^2 m_\pi^2(-1 + x + 2x^2)}}{1 + x}.$$

The first solution $s_1(x)$ gives always a cut along the negative real axis because the radicand is larger than the squared of the the terms in the numerator before the square root [their difference is $(1 - x^2)(m^2 - m_\pi^2)^2$]. Incidentally this also shows that the radicand is always positive for any values of the masses and $x \in [-1, 1]$. Its upper limit happens for $x = 1$ and it is zero, while its lower limit is $-\infty$ because $(1 + x)$ in the denominator tends to zero for $x \to -1$. This is a clear example of a left-hand cut (LHC). Regarding $s_2(x)$, this implies a finite cut which ranges along the positive real axis with values from $(m^2 - m_\pi^2)^2/m^2$ up to $m^2 + 2m_\pi^2$.

The analysis for $\pi\pi$ scattering is simpler because the two-pion cut along the t-channel and u-channel happens for $t = -2(s/4 - m_\pi^2)(1 - x) \geq 4m_\pi^2$ and $u = -2(s/4 - m_\pi^2)(1 + x) \geq 4m_\pi^2$. Solving s in terms of x we find that both cases give rise to LHCs with $s \in]-\infty, 0]$ when x moves along $[-1, 1]$.

In the case of a nonrelativistic theory, the quantum fields only involve annihilation operators (or creation ones for the Hermitian conjugate field, Chap. 5 of Ref. [3]) and crossing does not operate. Nonetheless, there is still a LHC in this case due to the particles exchanged that give rise to the potential. For instance, let us consider a Yukawa potential

$$V(r) = \alpha \frac{e^{-rm_\pi}}{r}. \tag{3.23}$$

Its Fourier transform is

$$V(\mathbf{q}^2) = \alpha \int d^3 r\, e^{-i\mathbf{q}\mathbf{r}} \frac{e^{-rm_\pi}}{r} = \frac{4\pi\alpha}{\mathbf{q}^2 + m_\pi^2}, \tag{3.24}$$

where $\mathbf{q} = \mathbf{p}' - \mathbf{p}$ is the three-momentum transfer. Its angular projection for particles without spin is simply

$$V_J(p, p') = \frac{1}{2} \int_{-1}^{+1} d\cos\theta \, V(\mathbf{q}^2) \, P_J(\cos\theta) \tag{3.25}$$

$$= -\frac{\pi\alpha}{pp'} \int_{-1}^{+1} d\cos\theta \frac{P_J(\cos\theta)}{\cos\theta - (p^2 + p'^2 + m_\pi^2)/(2pp')}.$$

The LHC results by the vanishing of the denominator. Thus, in order to study its appearance we isolate the term that produces it by proceeding as follows:

$$V_J(p, p') = -\frac{\pi\alpha}{pp'} \int_{-1}^{+1} d\cos\theta \frac{P_J(\cos\theta)}{\cos\theta - \xi} = -\frac{\pi\alpha}{pp'} \int_{-1}^{+1} d\cos\theta \frac{P_J(\cos\theta) - P_J(\xi)}{\cos\theta - \xi} \tag{3.26}$$

$$- \frac{\pi\alpha}{pp'} P_J(\xi) \int_{-1}^{+1} \frac{d\cos\theta}{\cos\theta - \xi},$$

with

$$\xi = \frac{p^2 + p'^2 + m_\pi^2}{2pp'}. \tag{3.27}$$

The term before the last one in Eq. (3.26) does not give rise to the LHC because when the denominator vanishes the numerator also does. Therefore, the last term is the only one responsible for the LHC and the integration over $\cos\theta$ can be done explicitly with the result

$$-\frac{\pi\alpha}{pp'} P_J(\xi) \int_{-1}^{+1} \frac{d\cos\theta}{\cos\theta - \xi} = -\frac{\pi\alpha}{pp'} P_J(\xi) \left[\log(1 - \xi) - \log(-1 - \xi)\right]. \tag{3.28}$$

Now for real and positive p and p' we can rewrite the difference of logarithms in the last term as

$$\frac{\pi\alpha}{pp'} P_J(\xi) \left[\log((p + p')^2 + m_\pi^2) - \log((p - p')^2 + m_\pi^2)\right]. \tag{3.29}$$

This expression is specially suitable for performing the analytical continuation to complex values of p and p' and so determine the position of the cuts, as fully exploited in Ref. [7]. The point is that the cuts in the p variable for given p' occurs when $(p + p')^2 + m_\pi^2 < 0$ (first logarithm) or $(p - p')^2 + m_\pi^2 < 0$ (second logarithm). This implies the vertical cuts

$$p = (\pm)p' \pm i\sqrt{m_\pi^2 + x^2}, \, x \in \mathbb{R}, \tag{3.30}$$

with the first $\pm$ symbol uncorrelated with the second one. An analogous reciprocal relation exists for the cuts in the variable p' for a given p. The cuts for on-shell scattering, $p = p'$, result from the only meaningful equation then, by taking the minus sign between brackets in Eq. (3.30),

$$p = -p \pm i \sqrt{m_\pi^2 + x^2}. \tag{3.31}$$

Its solution gives

$$p = \pm \frac{i}{2} \sqrt{m_\pi^2 + x^2}, \tag{3.32}$$

and for the variable p^2 we have a cut for the values

$$p^2 \leq -\frac{m_\pi^2}{4}. \tag{3.33}$$

This is the LHC that occurs in nonrelativistic nucleon–nucleon (NN) scattering [7]. The NN partial waves are function of the variable p^2 because by imposing parity invariance of the T matrix,

$$PTP = T, \tag{3.34}$$

in the equation that gives the projection onto the partial waves, Eq. (2.46), one easily deduces that

$$T_{ij}(-p) = \eta^2(-1)^{\ell_i + \ell_j} T_{ij}(p) = T_{ij}(p), \tag{3.35}$$

because $Y_\ell^m(-\hat{\mathbf{p}}) = (-1)^\ell Y_\ell^m(\mathbf{p})$ and $P|\mathbf{p}, \sigma_1\sigma_2, \alpha_1\alpha_2\rangle = \eta| - \mathbf{p}, \sigma_1\sigma_2, \alpha_1\alpha_2\rangle$, with η the intrinsic parity. We can write $\eta^2(-1)^{\ell_i + \ell_j} = +1$ because parity is a good quantum number and partial-wave states with different parity are not connected by time evolution.

Chapter 4
Important Mathematical Results: Schwarz Reflection Principle, Sugawara–Kanazawa Theorem, and Herglotz Theorem

• The *Schwarz reflection principle* states that given a function $f(z)$ of a complex variable z such that $f(z)$ is real in a finite segment Γ of the real axis, then

$$f(z) = f(z^*)^* \tag{4.1}$$

in a domain D in complex z plane if (i) $\Gamma \subset D$ and (ii) $f(z)$ is analytic in D.

This theorem intuitively follows if we think that the Taylor expansion of the analytic function $f(z)$ in D can be worked out in terms of the derivatives of this function at a point $z_0 \in \Gamma$ and evaluated along the real axis. As a result, since $f(z)$ is real in $z \in \Gamma$ it follows that the derivatives of any order at z_0 are real. Therefore, the Taylor expansion around z_0 satisfies Eq. (4.1). Since $f(z) = f(z^*)^*$ in this region, it is clear that its analytical continuation will also satisfy it since one proceeds by overlapping the new regions of analytical continuation with the previous ones where it is already satisfied.

The Schwarz reflection principle is important since it has many applications when writing down dispersion relations (DRs). For instance, it establishes that given a partial-wave amplitude $T(s)$ with a separation between the RHC and LHC, there is an interval along the real s axis in which $T(s)$ is real, and then this function satisfies the Schwarz reflection principle, Eq. (4.1). It also implies that the discontinuity of the function along any of the aforementioned cuts is $2i$ its imaginary part because then

$$f(x + i\varepsilon) - f(x - i\varepsilon) = f(x + i\varepsilon) - f(x + i\varepsilon)^* = 2i\Im f(x + i\varepsilon). \tag{4.2}$$

This principle could also be applied to an invariant part of a scattering amplitude $A(s, t, u)$ as a function of s with a t fixed that does not match with intermediate states in the t-channel, and such that it allows for a region in s without intermediate states in the s- and u-channels. Then $A(s, t, u)$ will satisfy the Schwarz reflection principle as a function of s (let us recall the on-shell relation that $s + t + u = \sum m_i^2$).

© The Author(s), under exclusive licence to Springer Nature Switzerland AG 2019
J. A. Oller, *A Brief Introduction to Dispersion Relations*, SpringerBriefs in Physics,
https://doi.org/10.1007/978-3-030-13582-9_4

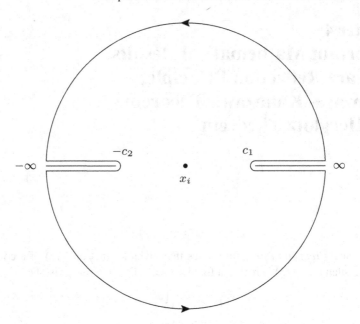

Fig. 4.1 Integration contour C used to settle the DR of Eq. (4.6)

• The *Sugawara–Kanazawa theorem* [15]: Let $f(z)$ be a function analytic everywhere in the complex z plane except for two cuts along the real axis and poles between them (as represented in Fig. 4.1). We also assume that

(i) $f(z)$ has finite limits $f(\infty \pm i\varepsilon)$ as $z \to \infty \pm i\varepsilon$ along the c_1 cut or RHC. ($\varepsilon \to 0^+$).
(ii) The possible divergence of $f(z)$ for $z \to \infty$ is less strong than a finite power $N \geq 1$ of z.
(iii) $f(z)$ has definite (not necessarily finite) limits for $z \to -\infty \pm i\varepsilon$ along the c_2 cut or LHC.[1]

The theorem then probes that $f(z)$ has the limits

$$\lim_{z \to \infty} f(z) = f(\infty + i\varepsilon), \quad \Im z > 0, \tag{4.3}$$

$$\lim_{z \to \infty} f(z) = f(\infty - i\varepsilon), \quad \Im z < 0,$$

and it admits the dispersion relation

$$f(z) = \sum_i \frac{R_i}{z - x_i} + \frac{1}{\pi} \left(\int_{c_1}^{\infty} + \int_{-\infty}^{-c_2} \right) \frac{\Delta f(x)}{x - z} dx + \bar{f}(\infty), \tag{4.4}$$

---

[1]The notation of Ref. [15] for a infinite definite limit refers to the fact that the function should tend to this infinite limit steadily without oscillating.

where

$$\Delta f(x) = \frac{1}{2i} [f(x + i\varepsilon) - f(x - i\varepsilon)], \tag{4.5}$$

$$\bar{f}(x) = \frac{1}{2} [f(x + i\varepsilon) + f(x - i\varepsilon)].$$

This theorem enables one to express the contribution from the infinite circle of the Cauchy contour integral in terms of the boundary values of $f(z)$ at infinity along only one of the cuts extending to infinity.

The Sugawara–Kanazawa theorem reflects a typical situation found in many applications, so that one has to study a function $f(z)$ with RHC and LHC that extend along the real axis up to infinity from some starting points, together with the possibility of poles along the real axis corresponding to bound states in the first Riemann sheet.

We offer here a summary of its demonstration, which is also of interest by itself since it provides the asymptotic behavior of integrals that are often found in actual applications. The demonstration also illustrates a peculiar manner of including extra subtractions in a DR.

We take the integration contour $\mathcal{C}$ in Fig. 4.1 and start with the usual integration for an unsubtracted DR,

$$\oint \frac{f(z')}{z' - z} dz' = \int_{c_1}^{R} \frac{f(x + i\varepsilon) - f(x - i\varepsilon)}{x - z} dx + \int_{-R}^{-c_2} \frac{f(x + i\varepsilon) - f(x - i\varepsilon)}{x - z} dx$$

$$+ \int_{circle} \frac{f(z')}{z' - z} dz' = 2\pi i f(z) + 2\pi i \sum_i \frac{R_i}{x_i - z}, \tag{4.6}$$

where R_i is the residue of $f(z)$ at the pole x_i. The set of poles x_i lies within the interval $[-c_2, c_1]$.

We take into account the asymptotic behaviors assumed in the statement of the theorem, so that we include one extra subtraction in the first integral on the rhs of Eq. (4.6), and $N + 1$ extra ones in the second and third integrals on the same side of the integral.

An interesting point in the demonstration, worth keeping in mind, is the way these extra subtractions are taken in Ref. [15] by employing the mathematical identity

$$\frac{1}{z' - z} = \sum_{i=0}^{m} \frac{z^i}{z'^{i+1}} + \frac{z^{m+1}}{z'^{m+1}} \frac{1}{z' - z}, \tag{4.7}$$

and applying it with $m = 0$ to the first integral on the rhs of Eq. (4.6) and with $m = N$ to the second and third integrals. The integrals along the circle of radius R involving $f(z')z^{N+1}/z'^{N+1}(z' - z)$ vanish in the limit $R \to \infty$. We can then rewrite $f(z)$ from Eq. (4.6) in the limit $R \to \infty$ as

$$f(z) = \sum_i \frac{R_i}{z - x_i} + \frac{z}{\pi} \int_{c_1}^{\infty} \frac{\Delta f(x)}{x(x - z)} dx + (-1)^N \frac{z^{N+1}}{\pi} \int_{c_2}^{\infty} \frac{\Delta f(-x)}{x^{N+1}(x + z)} dx$$

$$+ \left(a_0 + \frac{1}{\pi} \int_{c_1}^{\infty} \frac{\Delta f(x)}{x} dx - \frac{1}{\pi} \int_{c_2}^{\infty} \frac{\Delta f(-x)}{x} dx \right) \tag{4.8}$$

$$+ \left(a_1 + \frac{1}{\pi} \int_{c_2}^{\infty} \frac{\Delta f(-x)}{x^2} dx \right) z + \cdots + \left(a_N + \frac{(-1)^{N+1}}{\pi} \int_{c_2}^{\infty} \frac{\Delta f(-x)}{x^{N+1}} dx \right) z^N,$$

where we have exchange the sign of the integration variable in the integrals along the LHC. In this equation the coefficients a_m are the integrals along the circle at infinity from the last integral in Eq. (4.6) divided by $2\pi i$,

$$a_m = \lim_{R \to \infty} \frac{1}{2\pi R^m} \int_0^{2\pi} f(Re^{i\theta}) e^{-im\theta} d\theta. \tag{4.9}$$

By attending only to ii) many of the individual integrals in the coefficients of the polynomial in Eq. (4.8) are divergent, though their actual sum is convergent power by power in z, since both $f(z)$ and the first two integrals on the rhs of this equation are finite.

Now let us consider the constraints imposed by the condition i). Our analysis is in the following simpler than in the original Ref. [15], addressing faster the main reason why we proceed along with a sketch of the demonstration of the theorem. Those seeking more mathematical rigor are addressed to Ref. [15]. Specifically we discuss the first integral in Eq. (4.8) and rewrite it as

$$\frac{z}{\pi} \int_{c_1}^{\infty} \frac{\Delta f(x)}{x(x - z)} dx = \Delta f(\infty) \frac{z}{\pi} \int_{c_1}^{\infty} \frac{1}{x(x - z)} dx + \frac{z}{\pi} \int_{c_1}^{\infty} \frac{\Delta f(x) - \Delta f(\infty)}{x(x - z)} dx, \tag{4.10}$$

where $\Delta f(\pm\infty)$ is the limit

$$\Delta f(\pm\infty) = \lim_{x \to \pm\infty} \Delta f(x), \tag{4.11}$$

that exists and is finite in virtue of i). The first integral on the rhs of Eq. (4.10) as a function of z diverges logarithmically as $z \to \infty$ unless $\Delta f(\infty) = 0$, while the last integral has the limit

$$\lim_{z \to \infty} \frac{z}{\pi} \int_{c_1}^{\infty} \frac{\Delta f(x) - \Delta f(\infty)}{x(x - z)} dx = -\frac{1}{\pi} \int_{c_1}^{\infty} \frac{\Delta f(x) - \Delta f(\infty)}{x} dx. \tag{4.12}$$

Nonetheless, we cannot conclude that $\Delta f(\infty)$ is zero because still we have more contributions from the first LHC integral in Eq. (4.8) that could cancel such logarithmic divergent contribution from Eq. (4.10).

Indeed, we can proceed similarly for the first integral along the LHC in Eq. (4.8) and then it is required that[2]

$$\lim_{x \to -\infty} \frac{\Delta f(-x)}{x^N} = 0, \tag{4.13}$$

because otherwise we would have a contribution diverging like $z^N \log z$ in contradiction with i). Since Eq. (4.13) must be fulfilled then Eq. (4.12) implies that

$$\lim_{z \to \infty} (-1)^N \frac{z^{N+1}}{\pi} \int_{c_2}^{\infty} \frac{\Delta f(-x)}{x^{N+1}(x+z)} dx = \lim_{z \to \infty} (-1)^N \frac{z^N}{\pi} \int_{c_2}^{\infty} \frac{\Delta f(-x)}{x^{N+1}} dx, \tag{4.14}$$

which cancels the contribution from the integral along the LHC in the coefficient multiplying z^N. Similarly, $a_N = 0$ because otherwise we would have a x^N divergence at odds with (i). We can continue this analysis for $N - 1$. For this continuation let us notice that if Eq. (4.13) is fulfilled then we can rewrite

$$(-1)^N \frac{z^{N+1}}{\pi} \int_{c_2}^{\infty} \frac{\Delta f(-x)}{x^{N+1}(x+z)} dx = (-1)^N \frac{z^N}{\pi} \int_{c_2}^{\infty} \frac{\Delta f(-x)}{x^N} \frac{z+x-x}{x(x+z)} dx \tag{4.15}$$

$$= (-1)^{N-1} \frac{z^N}{\pi} \int_{c_2}^{\infty} \frac{\Delta f(-x)}{x^N(x+z)} dx + (-1)^N \frac{z^N}{\pi} \int_{c_2}^{\infty} \frac{\Delta f(-x)}{x^{N+1}} dx.$$

The term before the last one in the rhs of this equation is the same as the original one but with $N \to N - 1$, while the last factor is the same as already discussed in connection with Eq. (4.14). As a result, decreasing step by step in one unit the number of subtractions in the integral along the LHC, we can conclude that $a_m = 0$ and $\Delta f(-\infty)/x^m = 0$ for all $m \geq 1$. Therefore, it follows at this stage that only one subtraction is needed in Eq. (4.8) instead of the $N + 1$ taken there. The latter equation then simplifies to

$$f(z) = \sum_i \frac{R_i}{z - x_i} + \frac{z}{\pi} \int_{c_1}^{\infty} \frac{\Delta f(x)}{x(x-z)} dx + \frac{z}{\pi} \int_{c_2}^{\infty} \frac{\Delta f(-x)}{x(x+z)} dx \tag{4.16}$$

$$+ a_0 + \frac{1}{\pi} \int_{c_1}^{\infty} \frac{\Delta f(x)}{x} dx - \frac{1}{\pi} \int_{c_2}^{\infty} \frac{\Delta f(-x)}{x} dx.$$

We can proceed analogously as in the analysis of Eq. (4.10) and add and subtract $\Delta f(\pm\infty)$ in the first and second integrals in Eq. (4.16). This tells us that $f(x \pm i\varepsilon)$ would diverge for $x \to \infty$ unless

$$\Delta f(\infty) - \Delta f(-\infty) = 0. \tag{4.17}$$

---

[2]In the manipulations for the demonstration of this theorem we take that the vanishing limits for $x \to \infty$ tend to zero at least as $x^{-\gamma}$ with $\gamma > 0$.

By decomposing $1/x(x \pm z)$ in simpler fractions

$$\frac{1}{x(x \pm z)} = \mp \frac{1}{z} \left[\frac{1}{x \pm z} - \frac{1}{x} \right] \tag{4.18}$$

we can further rewrite Eq. (4.16) as

$$f(z) = \sum_i \frac{R_i}{z - x_i} + \frac{1}{\pi} \int_{c_1}^{\infty} \frac{\Delta f(x)}{x - z} dx - \frac{1}{\pi} \int_{c_2}^{\infty} \frac{\Delta f(-x)}{x + z} dx + a_0. \tag{4.19}$$

Notice that the integrals after a_0 in Eq. (4.16) cancel with the integrals that stem from the last term in Eq. (4.18). In general, it is convenient for considering the limit $z \to \infty$ of Eq. (4.19) to rewrite it as

$$f(z) = a_0 + \frac{1}{\pi} \int_{c_1}^{\infty} \Delta f(x) \left(\frac{1}{x - z} - \frac{1}{x + z} \right) dx - \int_{c_1 + c_2}^{\infty} \frac{\Delta f(-x) - \Delta f(x)}{x + z} dx + \cdots$$

$$= a_0 + \frac{\Delta f(\infty)}{\pi} \int_{c_1}^{\infty} \left(\frac{1}{x - z} - \frac{1}{x + z} \right) dx \tag{4.20}$$

$$+ \frac{1}{\pi} \int_{c_1}^{\infty} [\Delta f(x) - \Delta f(\infty)] \left(\frac{1}{x - z} - \frac{1}{x + z} \right) dx$$

$$- \int_{c_1 + c_2}^{\infty} \frac{\Delta f(-x) - \Delta f(x)}{x + z} dx + \cdots$$

where the ellipsis indicates contributions that trivially vanish as $z \to \infty$. The second and third integrals, the latter because of Eq. (4.17), are zero in the limit $z \to \infty$. The only surviving contribution stems from the first integral which gives

$$\lim_{z \to \infty} \frac{\Delta f(\infty)}{\pi} \int_{c_1}^{\infty} \left(\frac{1}{x - z} - \frac{1}{x + z} \right) dx = \begin{cases} +i \Delta f(\infty), & \Im z > 0, \\ -i \Delta f(\infty), & \Im z < 0. \end{cases} \tag{4.21}$$

This also fixes

$$a_0 = \bar{f}(\infty), \tag{4.22}$$

by taking the limit $z \to \infty \pm i\varepsilon$ in Eq. (4.20) which implies $f(\infty \pm i\varepsilon) = a_0 \pm i \Delta f(\infty)$. Thus, we conclude the validity of Eqs. (4.3) and (4.4).

Let us indicate some other important conclusions that follow from the Sugawara–Kanazawa theorem

1. If the RHC and LHC have finite extent, then the results of the theorem are trivial (the infinite point is isolated).
2. If the function only has one infinite cut, then $\Delta f(\infty) = 0$, because the function is continuous at the end of the opposite side of the real axis and then $\bar{f}(\infty) = f(\infty \pm i\varepsilon)$. This is the case for form factors.
3. If the Schwarz theorem is fulfilled, Eq. (4.1), then $\Delta f(x) = \Im f(x)$ and $\bar{f} = \Re f(\infty)$.

4. We have assumed that *both* limits $f(\infty \pm i\varepsilon)$ are finite in the statement of the theorem, point i). This is necessarily the case when one of them is finite and the Schwarz reflection principle is satisfied. Even in the case when only one of them is finite, let us say $f(\infty + i\infty)$, and the other $f(\infty - i\varepsilon)$ is infinite, the first line of Eq. (4.4) still holds. The demonstration is analogous to the one followed here, with an integration contour that consists of an infinite semicircle along the upper half of the complex z plane and a path along the entire real axis.

5. If $f(z)$ is known to approach zero as $z \to \infty \pm i\varepsilon$, the theorem states that $f(z)$ approaches zero in any other direction and the unsubtracted dispersion relation of Eq. (4.3) with $\bar{f}(\infty) = 0$ holds.

6. If $f(z)$ is known to diverge as z goes to either one or both of the limits $\infty \pm i\varepsilon$, we can introduce a function $F(z)$ which diverges at least as strongly and apply the theorem to the new function $f(z)/F(z)$. Of course, this new function should have finite limits for $\infty \pm i\varepsilon$ and cannot have poles and branch cuts out of the real axis, with the cuts lying separately along the latter. Since $F(z)$ is a known function we can calculate the residues and discontinuity of $f(z)/F(z)$ in terms of those of $f(z)$.

   As a result of applying the theorem to $f(z)/F(z)$ we conclude that $f(z)$ diverges at infinity as $F(z)$ times constants that are the limits of $f(z)/F(z)$ for $z \to \infty \pm i\varepsilon$.

• The *Herglotz theorem* [16]: Consider an analytic function $g(\xi)$ in $|\xi| < 1$ and which fulfills that $\Im g(\xi) \geq 0$ in this domain. Then, the function $g(\xi)$ can be represented by the integral

$$g(\xi) = i \int_0^{2\pi} \frac{e^{i\theta} + \xi}{e^{i\theta} - \xi} d\beta(\theta) + c, \qquad (4.23)$$

where $\beta(\theta)$ is an increasing bounded function and c is a real constant.

The transformation

$$\xi = \frac{z - i}{z + i}, \quad z = i\frac{1 + \xi}{1 - \xi}, \qquad (4.24)$$

maps the region $|\xi| < 1$ with the upper half plane $\Im z > 0$, while

$$x = i\frac{1 + e^{i\theta}}{1 - e^{i\theta}} = -\cot\frac{\theta}{2}, \qquad (4.25)$$

maps the circle $|\xi| = 1$ with the real z axis. Making use of this transformation one can easily rewrite the integral of Eq. (4.23) into the following integral representation of $f(z) = g(\xi)$ [16]:

$$f(z) = c + Az + \int_{-\infty}^{\infty} \frac{1 + xz}{x - z} d\alpha(x), \quad \Im z > 0, \qquad (4.26)$$

where c is the same constant as in Eq. (4.23), $\alpha(x) = \beta(\theta(x))$ (therefore, $\alpha(x)$ is an increasing bounded function in x) and A is another real constant.

It is also demonstrated in Ref. [16] that

$$\lim_{z \to \infty} \frac{f(z)}{z} = A, \ \Im z > 0. \tag{4.27}$$

Equation (4.26) provides too an analytical representation of $f(z)$ in the lower half plane $\Im z < 0$ such that it satisfies the Schwarz reflection principle, since c, A and $\alpha(x)$ are real. Thus,

$$f(z^*) = f(z)^*. \tag{4.28}$$

We can write the integral representation of Eq. (4.26) like a twice-subtracted DR by expressing the real constants c and A in terms of the function $f(z)$ and its derivative at some point z_0. In the following $f'(z)$ denotes the derivative of $f(z)$ at z. Making use of Eq. (4.26) we then have

$$f(z) = f(z_0) + f'(z_0)(z - z_0) + \frac{(z - z_0)^2}{\pi} \int_{-\infty}^{\infty} \frac{\sigma(x)}{(x - z_0)^2(x - z)} dx, \tag{4.29}$$

where

$$\sigma(x) = \pi(1 + x^2)\frac{d\alpha(x)}{dx} \geq 0. \tag{4.30}$$

We can also interpret $\sigma(x)$ as the imaginary part of $f(x + i\varepsilon)$ because from Eq. (4.26) one has that (recall that c and A are real)

$$\Im f(x \pm i\varepsilon) = \pm\pi(1 + x^2)\alpha'(x) = \pm\sigma(x). \tag{4.31}$$

This also has the consequence that we can write Eq. (4.29) as

$$f(z) = f(z_0) + f'(z_0)(z - z_0) + \frac{(z - z_0)^2}{\pi} \int_{-\infty}^{\infty} \frac{\Im f(x + i\varepsilon)}{(x - z_0)^2(x - z)} dx, \tag{4.32}$$

Note that Eq. (4.26) provides the valuable information that both c and A are real constants and that $\Im f(x + i\varepsilon) = \sigma(x)$, while these facts are not transparent in the form of the twice-subtracted DR in Eq. (4.32) for general z_0. Nonetheless, there is an exception to this comment that happens when $f(z)$ is real along some interval of the real axis because then it follows from Eq. (4.31) that $\sigma(x) = 0$ there. In such a case, we can take $z_0 = x_0$ along this interval and then both $f(x_0)$ and $f'(x_0)$ are real.

Chapter 5
Exact Dispersion Relations in Quantum Mechanics for the Eigenvalues of the Scattering Kernel

As an example for the application of the Sugawara–Kanazawa theorem and the Herglotz theorem introduced and discussed in Chap. 4, we derive the DR satisfied by the eigenvalues of the scattering kernel in Quantum Mechanics. The exposition is motivated by Ref. [17].

The Lippmann–Schwinger equation with a Hermitian potential V for a complex or negative energy W is[1]

$$T(W) = V + V[W - H_0]^{-1}T(W) , \qquad (5.1)$$

where $T(W)$ is the scattering matrix. We next make the transformation $V \to \lambda V$, with λ a complex number, and multiply both sides of Eq. (5.1) by $[W - H_0]^{-1}/\lambda$. Then we have

$$[W - H_0]^{-1}\frac{T(W)}{\lambda} = [W - H_0]^{-1}V + \lambda[W - H_0]^{-1}V[W - H_0]^{-1}\frac{T(W)}{\lambda} . \qquad (5.2)$$

This equation is the analogous one for finding the resolvent R of a Fredholm integral equation (IE) of the second type with kernel K [18]

$$R = K + \lambda K R . \qquad (5.3)$$

We can then make the following identifications:

$$K \to [W - H_0]^{-1}V , \qquad (5.4)$$

$$R \to [W - H_0]^{-1}\frac{T(W)}{\lambda} . \qquad (5.5)$$

---

[1]This equation differs by a minus sign compared to Eq. (2.65) because $V = -v$, so that $H = H_0 + V$.

© The Author(s), under exclusive licence to Springer Nature Switzerland AG 2019
J. A. Oller, *A Brief Introduction to Dispersion Relations*, SpringerBriefs in Physics,
https://doi.org/10.1007/978-3-030-13582-9_5

The values of λ that satisfy the homogeneous equation

$$[W - H_0]^{-1} V |\psi_\nu(W)\rangle = \eta_\nu(W) |\psi_\nu(W)\rangle \,, \tag{5.6}$$
$$\langle \psi_\nu(W) | \psi_\nu(W) \rangle = 1 \,,$$

are the eigenvalues of the interaction kernel. In order to connect with the theory of linear IE we should use $\eta_\nu(W)^{-1}$, because the eigenvalue equation is usually written as $\lambda K |\psi\rangle = |\psi\rangle$ [18].

It is also interesting to rewrite Eq. (5.6) as a more standard Schrödinger equation by multiplying it with $[W - H_0]$,

$$[H_0 + \frac{V}{\eta_\nu(W)}] |\psi_\nu\rangle = W |\psi_\nu\rangle \,. \tag{5.7}$$

In this way, if W is complex then $\eta_\nu(W)$ must be complex because the eigenvalues of a Hermitian Hamiltonian must be real.

We need some properties of the $\eta_\nu(W)$ as a function of W in order to show that they fulfill a DR of the type corresponding to the Herglotz theorem, Eq. (4.32).

(I) The first one follows upon inspection of Eq. (5.6); each $\eta_\nu(W)$ is analytic in the complex W plane cut along the real axis from 0 to ∞, because the spectrum of H_0 extends along real positive values. The only exception might arise if several $\eta_\nu(W)$ coalesce at the same value at some W. As in Ref. [17], we assume that it does not happen (we could take advantage of varying slightly the parameters in the theory if needed).

(II) The second property is that $\Im \eta_\nu(W)$ is always positive or negative in a given half of the complex W plane. We already remarked after Eq. (5.7) that $\Im \eta_\nu(W)$ cannot vanish for complex W because $\eta_\nu(W)$ must be necessarily complex. Therefore, we only have to determine the sign of $\eta_\nu(W)$ at some convenient point with positive or negative $\Im W$.

It follows from (I) and (II) that we can apply Eq. (4.32) from the Herglotz theorem either to $\eta_\nu(W)$ or to $-\eta_\nu(W)$, depending on weather $\Im \eta_\nu(W) \geq 0$ or ≤ 0 for $\Im W > 0$, respectively. At this stage, this implies a twice-subtracted DR for $\eta_\nu(W)$. We can go further by noticing two more properties of $\eta_\nu(W)$:

(III) The $\eta_\nu(W)$ are real for $W < 0$. This follows by multiplying the eigenvalue equation (5.6) by $[H_0 - W]^{1/2}$ so that

$$-[H_0 - W]^{-1/2} V [H_0 - W]^{-1/2} \left([H_0 - W]^{1/2} |\psi_\nu\rangle \right) = \eta_\nu(W) \left([H_0 - W]^{1/2} |\psi_\nu\rangle \right) \,. \tag{5.8}$$

Thus, the $\eta_\nu(W)$ are eigenvalues of the Hermitian operator

$$\tilde{K} = -[H_0 - W]^{-1/2} V [H_0 - W]^{-1/2} \tag{5.9}$$

and must be real for $W < 0$.

It is also clear that $\eta_\nu(W^*) = \eta(W)^*$ and they satisfy the Schwarz reflection principle.

(IV) The $\eta_\nu(W)$ vanish in the limit $W \to \infty$. This property is based on the fact that the trace of $\widetilde{K}\,\widetilde{K}^\dagger$ clearly vanishes for $W \to \infty$ ($\Im W \neq 0$ or $W < 0$) since it exists for any complex or negative W for a wide range of potentials. For detailed conditions on the potential the reader is referred to [17].

In virtue of (III) and (IV) we can rewrite Eq. (4.26) as an unsubtracted DR. First let us notice that (IV) implies that $A = 0$, cf. Eq. (4.27). Secondly, since $\Im \eta_\nu(\infty \pm i\varepsilon) = 0$ then, because of Eq. (4.31), we can rewrite the rest of Eq. (4.26) as

$$\eta_\nu(z) = c + \int_{-\infty}^{\infty} \frac{1 + t(z - t + t)}{t - z} d\alpha(t) \tag{5.10}$$

$$= c + \int_{-\infty}^{\infty} \frac{1 + t^2}{t - z} d\alpha(t) - \int_{-\infty}^{\infty} t \, d\alpha(t) . \tag{5.11}$$

When taking the limit $z \to \infty$ the integral before the last one is zero and then we are left with

$$c - \int_{-\infty}^{\infty} t \, d\alpha(t) = 0 , \tag{5.12}$$

because $\eta_\nu(z)$ is zero in the same limit. As a result, it follows from the Herglotz theorem that

$$\eta_\nu(W) = \frac{1}{\pi} \int_0^{\infty} \frac{\Im \eta_\nu(x + i\varepsilon)}{x - W} dx . \tag{5.13}$$

This dispersive representation for $\eta_\nu(W)$ could be also obtained by applying the Sugawara–Kanazawa theorem in virtue of the remark 2 when discussing such theorem.

Reference [17] shows the interest to study the eigenvalues $\eta_\nu(W)$ in order to settle a criterion for the validity of the Born series. The point is that the Neumann series for the resolvent $[W - H_0]^{-1} T(W)/\lambda$ can be applied as long as $|\lambda| < |\eta_\nu^{-1}|$ for all ν. This is a consequence of the standard Fredholm theory which establishes that the resolvent is a meromorphic function of λ. As a result if all $|\eta_\nu| < 1$ then the Born series is valid because the actual situation happens for $\lambda = 1$. Indeed, from the eigenvalue equation (5.6) we would have for the action of $T(W)$ on $|\psi_\nu(W)\rangle$,

$$T(W)|\psi_\nu(W)\rangle = \left(V + V[W - H_0]^{-1}V + V[W - H_0]^{-1}V[W - H_0]^{-1}V + \cdots\right)|\psi_\nu\rangle$$

$$= V \sum_{i=0}^{\infty} \eta_\nu(W)^i |\psi_\nu\rangle , \tag{5.14}$$

which is convergent for $|\eta_\nu(W)| < 1$. Indeed, this corresponds to the geometric series and one has that

$$T(W)|\psi_\nu(W)\rangle = \frac{V|\psi_\nu(W)\rangle}{1 - \eta_\nu(W)} . \tag{5.15}$$

An interesting condition of having obtained that $\eta_\nu(\infty + i\infty) = 0$ is that the Born series is always applicable at sufficiently high energy in potential scattering.

Chapter 6
General Results for Two-Meson Scattering in Partial Waves After Neglecting the Crossed-Channel Cuts. N/D Method

We discuss here the first phenomenological application in this work dedicated to the study performed in Ref. [19] of the lightest pseudoscalar–pseudoscalar scattering in S and P waves, with special emphasis in the analysis of the related spectroscopy. This study is based on the derivation of the formula which gives the general structure of the partial-wave amplitudes when the crossed-channel or unphysical cuts are neglected.

The unphysical cuts comprise two types of cuts in the complex s plane. For processes of the type $a + a \to a + a$, with $m_i = m_a$, there is only a LHC for $s < s_{\text{Left}}$. However for those ones of the type $a + b \to a + b$ with $m_1 = m_3 = m_a$ and $m_2 = m_4 = m_b$, apart from a LHC there is also a circular cut in the complex s plane for $|s| = m_2^2 - m_1^2$ [5], where we have taken $m_2 > m_1$. In the rest of this section, for simplicity in the formalism, we just refer to the LHC as if it comprises all the unphysical cuts. This is enough for our purposes in this section. In any case, had we worked in the complex p^2 plane all the cuts would be linear cuts and then only a LHC would be present for this variable. In such a case, the analysis would be mathematically analogous to the one presented here in the complex s plane.

Let us indicate by $T_L(s)$ a two-meson partial-wave amplitude with angular momentum L. We first analyze the elastic case and then we generalize the results for the coupled partial-wave amplitudes. In our chosen normalization, the RHC leads to the imaginary part of $T_L^{-1}(s)$ for s above threshold, $s_{\text{th}} = (m_1 + m_2)^2$, according to Eq. (2.51),

$$\Im T_L^{-1}(s) = -\frac{p}{8\pi\sqrt{s}} , \quad s \geq s_{\text{th}} , \tag{6.1}$$

where

$$p = |\mathbf{p}| = \frac{\sqrt{(s - (m_1 + m_2)^2)(s - (m_1 - m_2)^2)}}{2\sqrt{s}} \tag{6.2}$$

$$\equiv \frac{\lambda^{1/2}(s, m_1^2, m_2^2)}{2\sqrt{s}}$$

© The Author(s), under exclusive licence to Springer Nature Switzerland AG 2019
J. A. Oller, *A Brief Introduction to Dispersion Relations*, SpringerBriefs in Physics,
https://doi.org/10.1007/978-3-030-13582-9_6

is the CM three-momentum of the two-meson system.[1] The LHC, for $s < s_{\text{Left}}$, leads to

$$T_L(s + i\epsilon) - T_L(s - i\epsilon) = 2i\Im T_L(s) . \tag{6.3}$$

A way to obtain a $T_L(s)$ fulfilling Eqs. (6.1) and (6.3) is the N/D method [20], in which $T_L(s)$ is expressed as the quotient of two functions,

$$T_L(s) = \frac{N_L(s)}{D_L(s)} , \tag{6.4}$$

where the denominator function, $D_L(s)$, only has RHC and the numerator function, $N_L(s)$, contains the unphysical cuts but no RHC.

In order to take explicitly into account the behavior of a partial-wave amplitude near threshold, which vanishes like p^{2L}, we introduce the new function, T'_L, defined as

$$T'_L(s) = \frac{T_L(s)}{p^{2L}} , \tag{6.5}$$

which also satisfies relations of the type of Eqs. (6.1) and (6.3). So that we can write

$$T'_L(s) = \frac{N'_L(s)}{D'_L(s)} . \tag{6.6}$$

From Eqs. (6.1), (6.3) and (6.5), the functions $N'_L(s)$ and $D'_L(s)$ obey the equations,

$$\Im D'_L = Im T'^{-1}_L N'_L = -\rho(s)N'_L p^{2L} , \qquad\qquad s > s_{\text{th}} \tag{6.7}$$
$$\Im D'_L = 0 , \qquad\qquad s < s_{\text{th}}$$

$$\Im N'_L = \Im T'_L\, D'_L = \frac{\Im T_L}{p^{2L}} D'_L , \qquad\qquad s < s_{\text{Left}} \tag{6.8}$$

$$\Im N'_L = 0 . \qquad\qquad s > s_{\text{Left}}$$

Since N'_L and D'_L can be simultaneously multiplied by any arbitrary real analytic function without changing its ratio, T'_L, nor Eqs. (6.7) and (6.8), we choose the later as a polynomial made out of all the zeros of $N'_L(s)$. In this way, we can consider in the following that $N'_L(s)$ is free of poles and thus, the poles of a partial-wave amplitude correspond to the zeros of $D'_L(s)$.

Using dispersion relations for $D'_L(s)$ and $N'_L(s)$, we can write for these functions from Eqs. (6.7) and (6.8) that

$$D'_L(s) = -\frac{(s - s_0)^n}{\pi} \int_{s_{\text{th}}}^{\infty} ds' \frac{p(s')^{2L}\rho(s')N'_L(s')}{(s' - s)(s' - s_0)^n} + \sum_{m=0}^{n-1} \bar{a}_m s^m , \tag{6.9}$$

---

[1]The context makes clear when p is a four-momentum or the modulus of $\mathbf{p}$.

where n is the number of subtractions needed such that

$$\lim_{s \to \infty} \frac{N'_L(s)}{s^{n-L}} = 0 \qquad (6.10)$$

since, from Eq. (6.2)

$$\lim_{s \to \infty} \frac{p^{2L} \rho(s)}{s^L} = \frac{1}{4^{L+2} \pi} . \qquad (6.11)$$

For the function $N'_L(s)$ we have, consistently with Eq. (6.10),

$$N'_L(s) = \frac{(s-s_0)^{n-L}}{\pi} \int_{-\infty}^{s_{\text{Left}}} ds' \frac{\Im T_L(s') D'_L(s')}{p(s')^{2L}(s'-s_0)^{n-L}(s'-s)} + \sum_{m=0}^{n-L-1} \overline{a}'_m s^m . \qquad (6.12)$$

The Eqs. (6.9) and (6.12) constitute a system of coupled linear integral equations for the functions $N'_L(s)$ and $D'_L(s)$, which input is $\Im T_L(s)$ along the LHC.

However, Eqs. (6.9) and (6.12) are not the most general solution to Eqs. (6.7) and (6.8) because of the possible presence of zeros of T_L which do not originate when solving those equations. These zeros have to be included explicitly and we choose to incorporate them through poles in the function D'_L (the so-called Castillejo-Dalitz-Dyson (CDD) poles after Ref. [21]). Following this last reference, let us write along the real axis

$$\Im D'_L(s) = \frac{d\lambda(s)}{ds} . \qquad (6.13)$$

Then by Eq. (6.7),

$$\frac{d\lambda}{ds} = -\rho(s) p^{2L} N'_L , \qquad s > s_{\text{th}} \qquad (6.14)$$

$$\frac{d\lambda}{ds} = 0 . \qquad s < s_{\text{th}}$$

Let s_i be the points along the real axis where $T'_L(s_i) = 0$. Between two consecutive points, s_i and s_{i+1}, we have from Eq. (6.14) that

$$\lambda(s) = -\int_{s_i}^{s} p(s')^{2L} \rho(s') N'_L(s') ds' + \lambda(s_i) , \qquad (6.15)$$

with $\lambda(s_i)$ unknown because the inverse of $T'_L(s_i)$ is not defined. Thus, we may write

$$\lambda(s) = -\int_{s_{\text{th}}}^{s} p(s')^{2L} \rho(s') N'_L(s') ds' + \sum_i \lambda(s_i) \theta(s - s_i) , \qquad (6.16)$$

where $\theta(s)$ is the usual Heaviside function. Therefore, it follows from Eqs. (6.13) and (6.16) that

$$D'_L(s) = \frac{(s - s_0)^n}{\pi} \int_{S_{th}}^{\infty} \frac{\Im D'_L(s')ds'}{(s' - s)(s' - s_0)^n} + \sum_{m=0}^{n-1} \bar{a}_m s^m = \sum_{m=0}^{n-1} \bar{a}_m s^m \qquad (6.17)$$

$$- \frac{(s - s_0)^n}{\pi} \int_{S_{th}}^{\infty} \frac{p(s')^{2L} \rho(s') N'_L(s')}{(s' - s)(s' - s_0)^n} ds' + \frac{(s - s_0)^n}{\pi} \int_{S_{th}}^{\infty} \frac{\sum_i \lambda(s_i)\delta(s' - s_i)}{(s' - s)(s' - s_0)^n} ds'$$

$$= - \frac{(s - s_0)^n}{\pi} \int_{S_{th}}^{\infty} \frac{p(s')^{2L} \rho(s') N'_L(s')}{(s' - s)(s' - s_0)^n} ds' + \sum_{m=0}^{n-1} \bar{a}_m s^m + \sum_i \frac{\lambda(s_i)}{\pi(s_i - s)} \frac{(s - s_0)^n}{(s_i - s_0)^n} .$$

The Eq. (6.17) can also be obtained from Eq. (6.7) and the use of the Cauchy theorem for complex integration once the possible presence of the CDD poles of D'_L (zeros of T'_L) inside and along the integration contour are taken into account. The latter is given by a circle in the infinity deformed to engulf the real axis along the right-hand cut, $s_{th} < s' < \infty$. In this way one can also consider the possibility of there being higher order zeros and that some of the s_i could have a nonzero imaginary part (because of the Schwartz theorem, in this case s_i^* is another zero of $T'_L(s)$). However, as we see below for $L \leq 1$, when considering chiral symmetry in the Large N_c limit of Quantum Chromodynamics (QCD), the zeros will appear on the real axis and also as simple zeros. In general, by using T'_L instead of T_L, we avoid working with Lth order poles of D_L at threshold in the dispersion relation given in Eq. (6.17).

The last term in the rhs of Eq. (6.17) can also be written in a more convenient way by avoiding the presence of the subtraction point s_0. To accomplish this let us notice that

$$\frac{(s - s_0)^n}{s - s_i} = (s - s_0)^{n-1} \frac{s - s_i + s_i - s_0}{s - s_i} = (s - s_0)^{n-1} \left(1 + \frac{s_i - s_0}{s - s_i} \right) \qquad (6.18)$$

$$= (s - s_0)^{n-1} + (s_i - s_0) \frac{(s - s_0)^{n-1}}{s - s_i}$$

$$= \sum_{i=0}^{n-1} (s - s_0)^{n-1-i} (s_i - s_0)^i + \frac{(s_i - s_0)^n}{s - s_i} .$$

The terms

$$\sum_{i=0}^{n-1} (s - s_0)^{n-1-i} (s_i - s_0)^i \qquad (6.19)$$

can be reabsorbed in

$$\sum_{m=0}^{n-1} \overline{a}_m s^m \ . \tag{6.20}$$

As a result we can write Eq. (6.17) in the simpler manner

$$D'_L(s) = -\frac{(s-s_0)^n}{\pi} \int_{S_{th}}^{\infty} \frac{p(s')^{2L}\rho(s')N'_L(s')}{(s'-s)(s'-s_0)^n} ds' + \sum_{m=0}^{n-1} \widetilde{a}_m s^m + \sum_i \frac{\widetilde{\gamma}_i}{s-s_i} \ , \tag{6.21}$$

where $\widetilde{a}_m$ ($n-1 \geq m \geq 0$) and $\widetilde{\gamma}_i$, s_i ($i \geq 0$), are arbitrary parameters. However, if some of the s_i is complex there is another s_j such that $s_j = s_i^*$ and $\widetilde{\gamma}_j = \widetilde{\gamma}_i^*$, as we explained above. Every term in the last sum of Eq. (6.21) is referred to as a CDD pole, after Ref. [21].

The Eqs. (6.21) and (6.12) stand for the general integral equations for D'_L and N'_L, respectively. Next we make the approximation of neglecting the left-hand cut, that is, we set $\Im T_L(s) = 0$ in Eq. (6.12). Thus one has:

$$N'_L(s) = \sum_{m=0}^{n-L-1} \widetilde{a}'_m s^m \ . \tag{6.22}$$

As a result, $N'_L(s)$ is just a polynomial of degree $\leq n - L - 1$.[2] So we can write,

$$N'_L(s) = \mathcal{C} \prod_{j=1}^{n-L-1} (s - s_j) \ . \tag{6.23}$$

In Eq. (6.23) it is understood that if $n - L - 1$ is zero N'_L is just a constant. Thus, the only effect of N'_L, apart from the normalization constant $\mathcal{C}$, is the inclusion of, at most, $n - L - 1$ zeros in $T'_L(s)$. Nonetheless, we can always divide N'_L and D'_L by Eq. (6.23). The net result is that, when the LHC is neglected, it is always possible to take $N'_L(s) = 1$ and all the zeros of $T'_L(s)$ manifests as CDD poles of the denominator function. In this way,

$$T'_L(s) = \frac{1}{D'_L(s)} \ , \tag{6.24}$$

$$N'_L(s) = 1 \ ,$$

$$D'_L(s) = -\frac{(s-s_0)^{L+1}}{\pi} \int_{S_{th}}^{\infty} ds' \frac{p(s')^{2L}\rho(s')}{(s'-s)(s'-s_0)^{L+1}} + \sum_{m=0}^{L} a_m s^m + \sum_i^{M_L} \frac{R_i}{s-s_i} \ .$$

---

[2]One can always make that $n \geq L+1$ just by multiplying N'_L and D'_L by s^k with k large enough.

The number of free parameters present in Eq. (6.24) is $L + 1 + 2\varrho$, where ϱ is the number of CDD poles, M_L, minus the number of complex conjugate pairs of s_i. These free parameters have a clear physical interpretation. Consider first the term 2ϱ which comes from the presence of CDD poles in $D'_L(s)$, Eq. (6.24). In [22] the presence of CDD poles was linked to the possibility of there being elementary particles with the same quantum numbers as those of the partial-wave amplitude, that is, particles which are not originated from a given "potential" or exchange forces between the scattering states. One can think that given a $D'_L(s)$ we can add a CDD pole and adjust its two parameters in order to get a zero of the real part of the new $D'_L(s)$ with the right position and residue, having a resonance/bound state with the desired mass and coupling. In this way, the arbitrary parameters that come with a CDD pole can be related with the coupling constant and mass of the resulting particle. This is one possible interpretation of the presence of CDD poles. However, as we are going to see below, these poles can also enter just to ensure the presence of zeros required by the underlying theory, as the Adler zeros [23] for the S-wave meson-meson interaction in QCD. The derivative of the partial-wave amplitude at the zero fixes the other parameter of a CDD, $\widetilde{\gamma}_i$. With respect to the contribution $L + 1$ to the number of free parameters coming from the angular momentum L, it appears just because we have explicitly established the behavior of a partial-wave amplitude close to threshold, vanishing as p^{2L}. This is required by the centrifugal barrier effect, well known from Quantum Mechanics.

It should be stressed that Eq. (6.24) is the most general structure that an elastic PWA, with arbitrary L, has when the left- hand cut is neglected. The free parameters that appear there are fitted to the experiment or calculated from the basic underlying theory. In Ref. [19] the basic dynamics is expected to be QCD, but Eq. (6.24) could also be applied to other interactions apart from QCD, as the Electroweak Symmetry Breaking Sector [24] [which also has the symmetries [25] used to derive Eq. (6.24)].

We now give the necessary steps to generalize Eq. (6.24) to coupled channels by employing a matrix formalism. From the beginning we neglect the unphysical cuts. As a consequence $T_L(s)_{ij}$ will be proportional to $p_i^L p_j^L$. This makes that $T_L(s)_{ij}$, apart from the right-hand cut coming from unitarity (above the thresholds for channels i and j, s_{th}^i and s_{th}^j respectively), will have another cut for odd L between s_{th}^i and s_{th}^j due to the square roots present in p_i and p_j. This can be avoided by defining, in analogy with the elastic case Eq. (6.5), the matrix T'_L as

$$T'_L(s) = p^{-L} T_L(s) p^{-L} \,, \tag{6.25}$$

with p a diagonal matrix which elements are $p_{ij} = p_i \delta_{ij}$, where p_i is the modulus of the CM momentum of the channel i, $p_i = \dfrac{\lambda^{1/2}(s, m_{1i}^2, m_{2i}^2)}{2\sqrt{s}}$, with m_{1i} and m_{2i} the masses of the two mesons in channel i. In this way, T'_L has only the right-hand cut coming from unitarity and it is free of the cut associated with the product of $p_i^L p_j^L$ for odd L.[3]

Along the RHC the matrix $T'_L(s)$ satisfies

$$\Im T_L'^{-1}(s) = -p^L \rho(s) p^L = -\rho(s) p^{2L} , \tag{6.26}$$

where $\rho(s)$ is a diagonal matrix defined by

$$\rho(s) = -\frac{p}{8\pi\sqrt{s}}\theta(s) , \tag{6.27}$$

with $\theta(s)$ another diagonal matrix such that $\theta_i(s) = \theta(s - s_{\mathrm{th};i})$.

We write T'_L as a quotient of two matrices, N'_L and D'_L making use of the coupled channel version of the N/D method [26]

$$T'_L = D_L'^{-1} N'_L \tag{6.28}$$

We can always take N'_L free of poles and also containing all the zeros of T'_L. In such a case N'_L is just a matrix of polynomials in s of maximum degree $n - L - 1$, namely,

$$N'_L = Q_{n-L-1} . \tag{6.29}$$

Next, from Eqs. (6.26) and (6.28) one has

$$\Im D'_L(s) = -N'_L(s)\rho(s)p^{2L} , \tag{6.30}$$

that we employ to write the following DR for D'_L. It results

$$D'_L(s) = -\frac{(s - s_0)^n}{\pi} \int_0^\infty ds' \frac{Q_{n-L-1}(s')\rho(s')p^{2L}(s')}{(s' - s)(s' - s_0)^n} + P_{n-1} , \tag{6.31}$$

with P_{n-1} a matrix of polynomials of maximum degree $n - 1$.

Because N'_L is just a matrix of polynomials, it can be reabsorbed in D'_L to give rise to a new $\tilde{D}'_L$ which fulfills Eq. (6.30) but with $\tilde{N}'_L = 1$. In this way

---

[3] The formalism can be straightforwardly generalized to allow for different Ls of the initial and final states in the case of hadrons with different spins.

$$T'_L = \tilde{D}'^{-1}_L \tag{6.32}$$
$$\tilde{N}'_L = 1 \,,$$
$$\tilde{D}'_L = -\frac{(s - s_0)^{L+1}}{\pi} \int_0^\infty ds' \frac{\rho(s') p^{2L}(s')}{(s' - s)(s' - s_0)^{L+1}} + R(s) \,,$$

with $R(s)$ a matrix of rational functions whose poles contain the zeros of T'_L. This fact is in clear analogy with the role played by the CDD poles introduced above for the elastic scattering.

Chapter 7
Reaching the Unphysical Riemann Sheets. A Nonlinear Integral Equation to Calculate a PWA

Now, let us discuss how to proceed to calculate the T matrix of PWAs in an unphysical Riemann sheet (RS). In order to give a general discussion let us use a generic parameterization for a T matrix by explicitly isolating the RHC. Performing a DR of the inverse of the T matrix by employing Eq. 2.51 we have

$$T_L(s)^{-1} = \mathcal{N}_L(s)^{-1} + a(s_0) - \frac{s - s_0}{\pi} \int_{S_{th}} \frac{\rho(s')ds'}{(s' - s_0)(s' - s)} , \tag{7.1}$$

$$T_L(s) = \left[\mathcal{N}_L(s)^{-1} + g(s) \right]^{-1} . \tag{7.2}$$

where we have included a subtraction at s_0 because $\rho(s)$ tends to constant as $s \to \infty$. Here $\mathcal{N}_L(s)$ is a matrix that only has crossed-channel cut (although it could also have CDD poles). In the limit in which crossed cuts are neglected this function and the $N(s)$ function of the N/D method can be made to coincide. In addition, the dispersive integral plus the subtraction constant $a(s_0)$ (so that the result is independent of the subtraction point s_0) is denoted by $g(s)$. The matrices $g(s)$ and $a(s_0)$ are diagonal (recall that $\rho(s)$ is a diagonal matrix), whose matrix elements are explicitly,

$$g_i(s) = a_i(s_0) - \frac{s - s_0}{\pi} \int_{S_{th;i}}^{\infty} \frac{\rho_i(s')ds'}{(s' - s_0)(s' - s)} . \tag{7.3}$$

Notice that if the only singularities of $T_L(s)_{ij}$ were a RHC, a LHC, and possible poles in between the two cuts, and if it were furthermore bounded in the complex s plane by some power of s for $s \to \infty$, we could then apply the Sugawara–Kanazawa theorem, Chap. 4. This is clear because from unitarity we have that $T_{ij} = (S_{ij} - 1)/(2i\rho_i^{1/2}\rho_j^{1/2})$, $|S_{ij}| \le 1$, and we would expect for the case of finite-range interactions that S_{ij} tends to a definite limit for $s \to \infty + i\varepsilon$ (let us note that the Schwarz reflection principle is fulfilled by the PWA). We could then conclude from the application of the Sugawara–Kanazawa theorem that $T_{ij}(s)$ would tend to constant for $s \to \infty$, like $(S_{ij}(\infty + i\varepsilon) - 1)/(2i\rho_i(\infty + i\varepsilon)^{1/2}\rho_j(\infty + i\varepsilon)^{1/2})$ for

© The Author(s), under exclusive licence to Springer Nature Switzerland AG 2019
J. A. Oller, *A Brief Introduction to Dispersion Relations*, SpringerBriefs in Physics,
https://doi.org/10.1007/978-3-030-13582-9_7

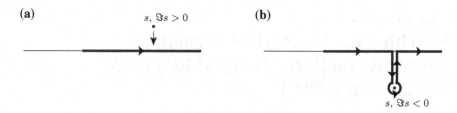

Fig. 7.1 Contour deformation (thick solid line) for reaching the second Riemann sheet of the function $g_i(s)$ by crossing the RHC from top (**a**) to bottom (**b**). The deformation of the integration contour results in order to avoid the pole singularity of the integrand in Eq. (7.3) at $s' = s$, for $s \in \mathbb{R}$ and $s > s_{\text{th};i}$. Subsequently the process continues to further avoid the crossing of the pole at s in the integrand with the deformed contour when moving deeper in the complex s plane. This figure could be seen also as a way to reach the first Riemann sheet from the second one by crossing again the RHC from the later (**a**) to the former (**b**). Of course, the RHC could also be crossed from bottom to top, with the deformed contour being the mirror image of the one pictured in (**b**)

$\Im s > 0$ and like its complex conjugate for $\Im s < 0$. Nonetheless, in practical applications we have to handle, at least at the effective level, with singular interactions for which the PWAs are not bounded in the complex s plane. For examples the interested reader might consult Ref. [7], where a formula is derived that allows to calculate the exact discontinuity of a PWA along the LHC both for regular and singular potentials. For the latter ones, the modulus of this discontinuity diverges stronger than any polynomial of s for $s \to -\infty$. Therefore, the Sugawara–Kanazawa theorem does not apply in this case and $T_{ij}(s)$ is divergent for $s \to \infty$, as the explicit calculation of the discontinuity along the LHC shows.

The Eq. (7.2) gives $T_L(s)$ in the first Riemann sheet. In order to reach resonance poles we should consider the T matrix in unphysical Riemann sheets as well. This is accomplished by performing the analytical continuation of the different matrix elements of the diagonal dispersive integral in Eq. (7.1). The function $g_i(s)$ has a branch-point singularity at the i_{th} threshold $s_{\text{th};i}$ and a cut starting from this point that we take along the positive real s axis, that is, a standard RHC or unitarity cut. Now, in order to reach the second Riemann sheet of $g_i(s)$ one should cross the RHC and proceed by analytical continuation to the second Riemann sheet. This analytical continuation can be accomplished by deforming the integration contour [2] as depicted in Fig. 7.1.

We then have to add to $g_i(s)$ the result of the integration along the closed integration contour around s. Thus, if we denote by $g_{II;i}(s)$ the $g_i(s)$ function in the second Riemann sheet we have the relation

$$g_{II;i}(s) = g_i(s) - 2i\rho_{II;i}(s) = g_i(s) + 2i\rho_{I;i}(s) \,, \tag{7.4}$$

where the function $\rho_{I;i}(s)$ in the complex s plane is

$$\rho_{I;i}(s) = \frac{1}{16\pi} \sqrt{\frac{\lambda(s, m_1^2, m_2^2)}{s^2}} , \qquad (7.5)$$

with the square root $\sqrt{z}$ taken in its first Riemann sheet and defined as having a RHC, that is, with $\arg z \in [0, 2\pi[$. Notice, that the minus sign in the term after the first equal sign in Eq. (7.4) is due to the fact that $\rho_{II;i}(s)$ is the same function as $\rho_{I;i}(s)$ but defined in its second RS (the procedure of analytically continuing an integral by deforming its integration contour requires using the integrand analytically continued to its corresponding Riemann sheet).

The Eq. (7.4) also shows that this is a two-sheet cut, because by crossing again the RHC we would have to add $+2i\rho_{I;i}(s)$, because of the addition of the circle to the integration contour, but this time added to $g_{II;i}(s) = g_i(s) + 2i\rho_{II;i}(s)$ (the square-root function $\rho_{I;i}(s)$ in Eq. (7.4) is also analytically continued to its second Riemann sheet). Then, the extra terms cancel and we come back again to $g_i(s)$ in the first Riemann sheet. This analysis shows that the RHC is a two-sheet cut and because of this the different Riemann sheets can be characterized as the Riemann sheets of the square root present in the definition of the CM three-momentum p,

$$p(s) = \pm \sqrt{\frac{\lambda(s, m_1^2, m_2^2)}{4s}} . \qquad (7.6)$$

Our convention to nominate all the possible 2^n RS for a scattering process with n channels is the following. The physical or first Riemann sheet (RS) corresponds to take the plus sign in all the channels, $(+, +, \ldots)$, the second RS to take the minus sign in the first channel, $(-, +, \ldots)$, the third RS to $(+, -, +, \ldots)$, the fourth RS to $(-, -, +, \ldots)$, etc. Thus, before we flip the sign of the m_{th} channel we have 2^{m-1} RSs.

We now discuss a DR for $\mathcal{N}_L(s)$ following the derivation of Ref. [8]. This representation also provides a nonlinear IE for $\mathcal{N}_L(s)$. To simplify the discussion we consider an uncoupled PWA taken as a function of the CM three-momentum squared, p^2. This is done so as to avoid the circular cuts for unequal mass scattering, §1.1 of Chap. 8 in Ref. [5], so that $T_L(p^2)$ has only a LHC and a RHC. The procedure discussed could be generalized straightforwardly to coupled PWAs.

From Eq. (7.2) we have that $\Im \mathcal{N}(p^2)$ satisfies along the LHC (we omit the subscript L to shorten the writing),

$$\Im T(p^2) = \Im \frac{1}{\mathcal{N}(p^2)^{-1} + g(p^2)} = -\frac{\Im \mathcal{N}(p^2)^{-1}}{|\mathcal{N}(p^2)^{-1} + g(p^2)|^2} = \Im \mathcal{N}(p^2) \frac{|T(p^2)|^2}{|\mathcal{N}(p^2)|^2} . \qquad (7.7)$$

Therefore,

$$\Im\mathcal{N}(p^2) = \frac{|\mathcal{N}(p^2)|^2}{|T(p^2)|^2}\Im T(p^2) \tag{7.8}$$

$$= \left|1 + g(p^2)\mathcal{N}(p^2)\right|^2 \Delta(p^2) , \quad p^2 < p^2_{\text{Left}} .$$

Here, we have introduced the function $\Delta(p^2)$ defined as

$$\Delta(p^2) = \Im T(p^2) , \quad p^2 < p^2_{\text{Left}} , \tag{7.9}$$

where p^2_{Left} is the upper bound of the LHC. Assuming that $\mathcal{N}(p^2)/p^{2n}$ vanishes for $p^2 \to \infty$ we can write an n-times subtracted DR for $\mathcal{N}(p^2)$,

$$\mathcal{N}(p^2) = \sum_{m=0}^{n-1} a_m p^{2m} + \frac{p^{2n}}{\pi} \int_{-\infty}^{p^2_{\text{Left}}} \frac{\left|1 + g(q^2)\mathcal{N}(q^2)\right|^2 \Delta(q^2)dq^2}{q^{2n}(q^2 - p^2)} . \tag{7.10}$$

This is a nonlinear IE which input is the knowledge of $\Delta(p^2)$ along the LHC. In terms of this DR we can write for $T(p^2)$,

$$T(p^2) = \left[\left(\sum_{m=0}^{n-1} a_m p^{2m} + \frac{p^{2n}}{\pi} \int_{-\infty}^{p^2_{\text{Left}}} \frac{\left|1 + g(q^2)\mathcal{N}(q^2)\right|^2 \Delta(q^2)dq^2}{q^{2n}(q^2 - p^2)}\right)^{-1} + g(p^2)\right]^{-1} . \tag{7.11}$$

In this form the subtraction constants a_m can be determined in terms of physical parameters of the T matrix, e.g., by fitting phase shifts, reproducing the effective range expansion (ERE) shape parameters, etc.

The Ref. [8] also shows that $T(p^2)$ is independent of the subtraction constant in $g(s)$. We reproduce here the arguments given in this reference and, as there, we take only one subtraction constant in $\mathcal{N}(p^2)$, which is enough for illustrating the point. We perform a DR for $T^{-1}(p^2)$ taking into account the RHC and LHC with an integration contour that consists of a circle at infinity that engulfs the two mentioned cuts. We use that the $\Im T(p^2)^{-1}$ along the RHC is $-\rho(p^2)$, Eq. 2.51. Then, one has

$$T^{-1}(p^2) = \beta - \frac{p^2}{\pi} \int_0^{\infty} \frac{\rho(q^2)dq^2}{q^2(q^2 - p^2)} + \frac{p^2}{\pi} \int_{-\infty}^{p^2_{\text{Left}}} \frac{\Delta(q^2)dq^2}{|T(q^2)|^2 q^2(q^2 - p^2)} + R(p^2) , \tag{7.12}$$

where $R(p^2)$ is a rational function taking care of the possible zeroes of $T(p^2)$ and that it does not play an active role in the considerations that follow. It is clear from the previous equation that there is only one free parameter (subtraction constant) to be determined, β, even though we could split it in two constants and add one of them to the integral over the RHC. The sum of this constant plus the RHC integral is the unitarity function $g(s)$. Thus, the inclusion of a subtraction constant in $g(p^2)$ appears just as a matter of convenience.

Chapter 8
The Good (σ), The Bad (ρ) and the Difficult (LHC)

We now apply the formalism developed in Chap. 6 to study the low-energy isoscalar S-wave and isovector P-wave $\pi\pi$ amplitudes. Their phase shifts are characterized by the presence of a broad increase and later plateau (σ) for the former and by a steep rise (ρ) for the later.

As dynamical input we first consider the lowest order Chiral Perturbation Theory (ChPT) amplitudes. As it is well-known Chiral Perturbation Theory is a low-energy effective field theory (EFT) employing as effective degrees of freedom the lightest pseudoscalars that correspond to the pseudo-Goldstone bosons associated with the spontaneous chiral symmetry breaking of strong interactions. Another consequence of the Goldstone theorem is that the interaction involving the Goldstone bosons are of derivative nature and vanish in the limit $p_i^2 \to 0$. As a result, even for the S waves there is a near threshold zero that is known as Adler zero, while for the P wave this is just the zero at threshold. For detailed accounts on this EFT the interested reader can consult the Chap. 19 of Refs. [27, 28] or the topical reviews [29–31].

We can treat both types of zeroes similarly by including a CDD pole in the corresponding $D(s)$ function. The leading order (LO) ChPT amplitudes are ($f_\pi \simeq 92.4$ MeV is the weak pion decay constant)

$$V_0 = \frac{s - m_\pi^2/2}{f_\pi^2} , \tag{8.1}$$

$$V_1 = \frac{s - 4m_\pi^2}{6f_\pi^2} ,$$

for the $I = 0$ S-wave and $I = 1$ P-wave $\pi\pi$ scattering, in order [19]. Thus, by approaching the position of the zero and its residue by LO ChPT, and by using Eq. (6.24) with $L = 0$ and $M_L = 1$, we have

J. A. Oller, *A Brief Introduction to Dispersion Relations*, SpringerBriefs in Physics, https://doi.org/10.1007/978-3-030-13582-9_8

$$T_0(s) = \left(\frac{f_\pi^2}{s - m_\pi^2/2} + \tilde{a}_0 - \frac{s}{\pi} \int_{S_{th}}^\infty ds' \frac{\rho(s')}{s'(s'-s)} \right)^{-1}, \qquad (8.2)$$

$$T_1(s) = \left(\frac{6 f_\pi^2}{s - 4 m_\pi^2} + \tilde{a}_1 - \frac{s}{\pi} \int_{S_{th}}^\infty ds' \frac{\rho(s')}{s'(s'-s)} \right)^{-1}.$$

In order to continue forward, and appreciate the difference between the σ and the ρ from the hadronic point of view of taking pions as explicit degree of freedom, we have to elaborate on the physical meaning of the subtraction constants $\tilde{a}_I$ in Eq. (8.2), which become part of the functions $g_I(s)$. These subtraction constants appear together with the dispersive integral so that the result is independent of the subtraction point s_0, which in Eq. (8.2) is equal to zero.

Let us first give the generic algebraic expression for the unitary function $g(s)$, cf. Eq. (7.3), with two particles involved in the intermediate state with masses m_1 and m_2 (in the scattering of a heavy particle with a much lighter one we take m_1 the mass of the heavier one),

$$g(s) = \tilde{a} - \frac{s}{\pi} \int_{S_{th}}^\infty ds' \frac{\rho(s')}{s'(s'-s)} \qquad (8.3)$$

$$= \frac{1}{16\pi^2} \left[a(\mu) + \log \frac{m_1^2}{\mu^2} - x_+ \log \frac{x_+ - 1}{x_+} - x_- \log \frac{x_- - 1}{x_-} \right],$$

$$x_\pm = \frac{s + m_2^2 - m_1^2}{2s} \pm \frac{1}{2s} \sqrt{(s + m_2^2 - m_1^2)^2 - 4s(m_2^2 - i0^+)}.$$

Here the renormalization scale μ is introduced to make dimensionless the argument of the first logarithm. The combination $a(\mu) - 2 \log \mu$ is independent of μ. The relation between $g(s_{th})$ and $a(\mu)$ is

$$g(s_{th}) = \frac{a(\mu)}{16\pi^2} + \frac{1}{8\pi^2(m_1 + m_2)} (m_1 \log \frac{m_1}{\mu} + m_2 \log \frac{m_2}{\mu}). \qquad (8.4)$$

The unitarity function $g(s)$ corresponds to the unitarity loop function

$$g(s) = i \int \frac{d^4 p}{(2\pi)^4} \frac{1}{[(P/2 - p)^2 - m_1^2 + i\varepsilon][(P/2 + p)^2 - m_2^2 + i\varepsilon]} \qquad (8.5)$$

$$= \int_0^\infty \frac{p^2 dp}{(2\pi)^2 \omega_1 \omega_2} \frac{\omega_1 + \omega_2}{[s - (\omega_1 + \omega_2)^2 + i\varepsilon]},$$

where $P = p_1 + p_2$ is the total four-momentum and $\omega_i = \sqrt{m_i^2 + \mathbf{p}^2}$. This integral is logarithmically divergent and requires regularization. This is why a subtraction constant is needed in Eq. (8.3). This equation can also be obtained by performing the loop integral with dimensional regularization in d dimensions and then absorbing in the subtraction constant the diverging contribution $1/(d-4)$ in the limit $d \to 4$. One could use as well a three-momentum cutoff regularization by integrating over p

up to a maximum Λ, and the resulting expression for $g(s)$ is worked out in Ref. [32] with the result

$$
g_A(s) = \frac{1}{32\pi^2} \left(-\frac{\Delta}{s} \log \frac{m_1^2}{m_2^2} + 2\frac{\Delta}{s} \log \frac{1+\sqrt{1+m_1^2/\Lambda^2}}{1+\sqrt{1+m_2^2/\Lambda^2}} + \log \frac{m_1^2 m_2^2}{\Lambda^4} \right. \tag{8.6}
$$

$$
+ \frac{\nu}{s} \left\{ \log \frac{s - \Delta + \nu\sqrt{1+m_1^2/\Lambda^2}}{-s + \Delta + \nu\sqrt{1+m_1^2/\Lambda^2}} + \log \frac{s + \Delta + \nu\sqrt{1+m_2^2/\Lambda^2}}{-s - \Delta + \nu\sqrt{1+m_2^2/\Lambda^2}} \right\}
$$

$$
\left. - 2 \log \left[\left(1 + \sqrt{1+m_1^2/\Lambda^2}\right) \left(1 + \sqrt{1+m_2^2/\Lambda^2}\right) \right] \right),
$$

with $\nu = \lambda(s, m_1^2, m_2^2)^{1/2}$ and $\Delta = m_2^2 - m_1^2$.

In hadron physics the natural size for a three-momentum cutoff is the size of the hadron themselves as resulting from the strong dynamics binding of quarks and gluons, the degrees of freedom of QCD. This value is around 1 GeV, since for a three-momentum cutoff larger than this value the associated de Broglie wave length would be smaller than the distances inside which extra degrees of freedom are manifest. The relativistic limit of both functions $g(s)$ and $g_A(s)$ ($|\mathbf{p}| \ll m_1, m_2$) is a constant plus $-ip/(8\pi(m_1 + m_2)) + \mathcal{O}(p^2)$. This constant for $g(s)$ is already given in Eq. (8.4), while for $g_A(s_{\text{th}})$ is

$$
g_A(s_{\text{th}}) = -\frac{1}{8\pi^2(m_1 + m_2)} \left[m_1 \log \left(1 + \sqrt{1+m_1^2/\Lambda^2}\right) \right. \tag{8.7}
$$

$$
\left. + m_2 \log \left(1 + \sqrt{1+m_2^2/\Lambda^2}\right) - m_1 \log \frac{m_1}{\Lambda} - m_2 \log \frac{m_2}{\Lambda} \right].
$$

We can then obtain a prediction for the constant $a(\mu)$ as a function of Λ by equating Eqs. (8.4) and (8.7)

$$
a(\mu) = -\frac{2}{m_1 + m_2} \left[m_1 \log \left(1 + \sqrt{1+m_1^2/\Lambda^2}\right) \right. \tag{8.8}
$$

$$
\left. + m_2 \log \left(1 + \sqrt{1+m_2^2/\Lambda^2}\right) \right] - \log \frac{\Lambda^2}{\mu^2}.
$$

For a given Λ we can obtain any value of $a(\mu)$ by varying μ. For instance for $\pi\pi$ scattering with $\Lambda = 1$ GeV we obtain

$$
a(\mu) = -1.40 - \log \frac{\Lambda^2}{\mu^2}, \quad \Lambda = 1 \text{ GeV}. \tag{8.9}
$$

By taking in Eq. (8.8) the renormalization scale to be equal to Λ and the latter to stay around 1 GeV, we obtain what is called the natural value for the subtraction constant, a concept introduced in Ref. [33]. Thus, taking this value for $a(\mu)$ in the isoscalar scalar $\pi\pi$ partial-wave amplitude we find a pole for the resonance σ or $f_0(500)$ at[1]

$$s_\sigma = (0.47 - i\, 0.20)^2\,, \tag{8.10}$$

which is compatible with the values given in Particle Data Group (PDG) [34] that reports

$$s_\sigma = (0.4 - 0.5 - i\,(0.20 - 0.35))^2 \text{ GeV}^2. \tag{8.11}$$

Then, it is clear that the σ resonance has a dynamical origin which finds an explanation within simple terms when employing the pions as explicit degrees of freedom. It is due to an interplay between the low-energy Adler zero located around $m_\pi^2/2$ (there could be small higher order corrections) and the rescattering of the two pions when propagating. For a recent and comprehensive review on the σ resonance the reader can consult Ref. [35].

However, if we apply the same idea to the isovector vector $\pi\pi$ PWA we cannot obtain the pole of the $\rho(770)$ with $a(\mu)$ having a natural value. We need to use a much larger subtraction constant in absolute value in order to obtain a good pole for the $\rho(770)$. For instance, for $a(\mu) = -14$, $\mu = 1$ GeV, we obtain

$$s_\rho = (0.777 - i\, 0.072)^2 \text{ GeV}^2\,, \quad a(1 \text{ GeV}) = -14\,. \tag{8.12}$$

This value of $a(\mu)$ requires an extremely big cutoff of around $\Lambda = 600$ GeV, near the TeV region. Thus, the presence of the $\rho(770)$ cannot be easily explained employing the pions as explicit degrees of freedom and the parameters of the theory need to be adjusted. This clearly indicates that this resonance has a nature very different to that of the σ.[2]

The Ref. [19] matched the rational function in $D(s)$ so as to include an explicit bare ρ resonance. In this way, it is clear that the large nonnatural value of the subtraction constant $a(\mu)$ for the $\rho(770)$ resonance is due to the fact that it is not a $\pi\pi$ rescattering effect but an elementary (quark–antiquark) resonance.

$$N_1(s) = 1 \tag{8.13}$$

$$D_1(s) = \sum_{i=1}^{2} \frac{\gamma_i}{s - s_i} + a - \frac{s - s_0}{\pi} \int_{s_{th}}^{\infty} ds' \frac{\rho(s')}{(s' - s)(s' - s_0)}\,,$$

---

[1] The pole is located in the 2nd Riemann sheet. We discuss above in Chap. 7 how to change to different Riemann sheets.

[2] In hadron physics an appealing justification for the appearance of the $\rho(770)$ in the spectrum is obtained by gauging the chiral symmetry in the nonlinear chiral Lagrangians [36–38].

with one of the CDD poles corresponding to the zero at threshold, already accounted for in $T_1(s)$ of Eq. (8.2). The tree-level LO ChPT amplitude plus the bare exchange of a ρ resonance employing the Lagrangian of Ref. [39] can be easily calculated and it gives [19]

$$t_1(s) = \frac{2}{3}\frac{p^2}{f_\pi^2}\left[1 + g_v^2\frac{s}{M_\rho^2 - s}\right] . \tag{8.14}$$

The KSFR [40] relation requires the coupling g_v to be equal to one. We can match this tree-level amplitude with two CDD poles. The position of a zero corresponds to that of the CDD pole and the derivative of $t_1(s)$ at this zero is the inverse of the residue of the CDD pole in the $D(s)$ function. The new CDD pole location and its residue corresponding to $t_1(s)$ are

$$s_2 = \frac{M_\rho^2}{1 - g_V^2} , \tag{8.15}$$

$$\gamma_2 = \frac{6f_\pi^2}{1 - g_v^2}\frac{g_v^2 M_\rho^2}{M_\rho^2 - 4(1 - g_v^2)m_\pi^2} .$$

In the limit $g_v^2 \to 1$ the zero s_2 tends to infinity and this exemplifies why we could generate before the ρ resonance by only a subtraction constant added to $g(s)$ in Eq. (8.2). Indeed the limit

$$\lim_{s_2 \to \infty} \frac{\gamma_2}{s - s_2} = -\frac{6f_\pi^2}{M_\rho^2} . \tag{8.16}$$

times $16\pi^2$ gives

$$-\frac{96\pi^2 f_\pi^2}{M_\rho^2} = -13.6 , \tag{8.17}$$

which is our value above for $a(1\text{ GeV})$ in the isovector vector $\pi\pi$ scattering. Therefore, this number reflects the elementary nature of the ρ resonance from the point of view of pionic degrees of freedom. In summary,

$$T_1(s) = \left[\frac{6f_\pi^2}{s - 4m_\pi^2} - \frac{6f_\pi^2}{M_\rho^2} + \frac{1}{16\pi^2}\left(\log\frac{m_\pi^2}{\mu^2} - x_+\log\frac{x_+ - 1}{x_+} - x_-\log\frac{x_- - 1}{x_-}\right)\right]^{-1} , \tag{8.18}$$

with $\mu \simeq 1$ GeV. A similar equation also holds for the $I = 1/2$ vector $K\pi$ scattering and the $K^*(892)$ resonance [19].

The Ref. [19] also includes explicit bare-resonance fields for the scalar sector and employs the matrix notation of Eq. (6.32) for coupled channels in order to perform a simultaneous study of the phase shifts and inelasticity parameters for $I = 0$ ($\pi\pi$, $K\bar{K}$ and $\eta_8\eta_8$), 1 ($\pi\eta_8$ and $K\bar{K}$) and 1/2 ($K\pi$ and $K\eta_8$), where we have singled out between brackets the coupled channels in every case. The unitarized amplitude reads for each isospin

$$T_I(s) = \left[t_I(s)^{-1} + g(s)\right]^{-1} , \qquad (8.19)$$

with $t_I(s)$ the tree-level amplitudes calculated from LO ChPT and the chiral Lagrangians of Ref. [39] employed for including explicit bare-resonance scalar fields. For instance, the tree-level amplitudes for $I = 1$ read

$$t_{1;11}(s) = \frac{m_\pi^2}{3f_\pi^2} + \frac{\beta_1^2}{M_S^2 - s} , \qquad (8.20)$$

$$t_{1;12}(s) = -\frac{\sqrt{3/2}}{12f_\pi^2}(6s - 8m_K^2) + \frac{\beta_1\beta_2}{M_S^2 - s} ,$$

$$t_{1;22}(s) = \frac{s}{4f_\pi^2} + \frac{\beta_2^2}{M_S^2 - s} ,$$

where the bare-resonance couplings β_i are

$$\beta_1 = \frac{\sqrt{2}}{\sqrt{3}f_\pi^2}\left(c_d(s - m_\pi^2 - m_\eta^2) + 2c_m m_\pi^2\right) , \qquad (8.21)$$

$$\beta_2 = -\frac{2}{f_\pi^2}\left(c_d\frac{s}{2} + (c_m - c_d)m_K^2\right) ,$$

here m_K is the kaon mass and $m_\eta^2 = 4m_K^2/3 - m_\pi^2/3$ (close to the eta mass). These tree-level amplitudes, as well as the ones used for the other channels, can be found in Ref. [19].

The function $g(s)$ is the unitarity loop function given in Eq. (8.3) and the subtraction constant a is taken to be the same for all channels and it is fitted in Ref. [19] with a resulting value of around -0.7 [19]. We discuss below in Chap. 9 that implementing exact $SU(3)$ symmetry requires the same value for the subtraction constants of all channels that are $SU(3)$ related. In this way, together with the σ resonance the study of Ref. [19] also obtains poles corresponding to the light scalar resonances, $f_0(980)$ ($I = 0$), $a_0(980)$ ($I = 1$), κ ($I = 1/2$), and to an octet of scalar resonances with masses near to 1.4 GeV with $I = 0, 1/2$, and 1. This mass is coincident with the mass of the bare $SU(3)$ octet of scalar resonances introduced in the evaluation of the tree-level amplitudes. The Ref. [19] also includes a singlet bare resonance with a mass around 1 GeV. This contribution to the $f_0(980)$ resonance is necessary so as to reproduce properly the inelasticity parameter η_1 for the isoscalar scalar S matrix, cf. Eq. (2.54), once the $\eta_8\eta_8$ is included as a one more coupled channel. This result answers the question raised in Ref. [41] about the unsatisfactory reproduction of the inelasticity parameter for the unitarized isoscalar scalar ChPT tree-level amplitudes after included the $\eta_8\eta_8$ channel in addition to the $\pi\pi$ and $K\bar{K}$ ones. Additionally, the results of Ref. [19] provide a fine reproduction of the experimental phase shifts and inelasticities for the different channels, cf. Figs. 2–8 in this reference.

All these results are obtained without having included explicitly LHC contributions. The latter are estimated in Ref. [19] and it is concluded that they are small in the resonant scalar meson–meson channels for the region $\sqrt{s} \lesssim 1$ GeV. The estimate is performed by considering the crossed loop diagrams that occur at $\mathcal{O}(p^4)$, or next-to-leading order (NLO), in the ChPT meson–meson amplitudes, and by including additionally the t- and u-channel exchanges of resonances with spin≤ 1. These contributions, denoted by $T_{I;\mathrm{Left}}(s)$, can be obtained from the ChPT calculation in Ref. [42] of the meson–meson scattering amplitudes with some bare resonances included. Being more precise, we match the calculated PWA $A(s)$ from Ref. [42] with the expansion of Eq. (8.19) at the one-loop level,

$$t_I(s) - t_I(s)g(s)t_I(s) , \tag{8.22}$$

from where $T_{I;\mathrm{Left}}(s)$ is defined as

$$T_{I;\mathrm{Left}}(s) = A(s) - t_I(s) + t_I(s)g(s)t_I(s) . \tag{8.23}$$

In other terms, we remove from $A(s)$ the tree-level amplitude and the unitarity contribution at the one-loop level, which is already taken into account by employing $T_I(s)$.[3] It follows then from the results of Ref. [19] that for the $L = 0, I = 0$ $\pi\pi$ and $L = 0, I = 1/2$ $K\pi$ elastic scattering (the lightest channel for every given quantum number is selected), the absolute value of the ratio $T_{I;\mathrm{Left}}(s)/t_I(s)$ is always $\lesssim 5\%$ up to around $\sqrt{s} = 1$ GeV. This smallness of the LHC contribution is remarkable and is due to the cancelation to a large extent between the crossed exchange of resonances in the t and u channels and the crossed loops (each of them separately is around a 15% of $t_I(s)$ in the considered region).

It is also wroth stressing that the previous cancelation between resonance exchanges and loops in the crossed channels also signals towards an important fact in the meson–meson scalar sector, which is the violation of large N_c QCD expectations. Notice that meson–meson loops are subleading in the large N_c

counting, while resonance exchanges are expected to be leading in this counting [43]. In connection with large N_c QCD, Ref. [19] was the first study in the literature to point out that the mass of the σ resonance does not follow the standard pattern for a $q\bar{q}$ resonance, so that it raises with N_c instead of being $\mathcal{O}(N_c^0)$. This point can be

---

[3]The Ref. [19] does not include in $A(s)$ the tadpole contributions for calculating $T_{I;\mathrm{Left}}(s)$. The idea followed by Ref. [19] is to keep only those contributions in the calculation of Ref. [42] that involve explicit LHC contributions, while the tadpoles are local ones. These tadpole contributions could also be removed by employing a different regularization scheme for the calculation.

understood from the expressions of $T_0(s)$ in Eq. (8.2) and $a(\mu)$ in Eq. (8.8), which show that the latter is $\mathcal{O}(N_c^0)$ as the rest of terms in $g(s)$. Therefore, when looking for a zero in the denominator of Eq. (8.2) we then have an equation for s of the type

$$s_\sigma \rightarrow - f_\pi^2/g(s_\sigma) = \mathcal{O}(N_c) , \qquad (8.24)$$

since f_π^2 runs like N_c in the large N_c QCD counting [43]. This situation should be compared with that for the ρ resonance as follows by employing $T_1(s)$ from Eq. 8.2 and the expression for $a(\mu)$ in Eq. (8.17) (notice that $\tilde{a}_1 = a(\mu)/16\pi^2$). For this case the equation that results is

$$s_\rho \rightarrow M_\rho^2 + \mathcal{O}(N_c^{-1}) , \qquad (8.25)$$

which counts as $\mathcal{O}(N_c^0)$. These manipulation also clearly show the very different nature regarding the origin of the σ and ρ resonances.

Two more interesting facts are also explored in a novel way in Ref. [19]. The first one is to notice that the set of the lightest scalar resonances [σ, κ, $f_0(980)$ and $a_0(980)$] have pole positions that largely vary as the pseudoscalar masses do. The second finding is that these resonances merge together in an octet plus a singlet in the chiral $SU(3)$ limit (this is the limit in which all the pseudoscalar masses vanish). That is, out of the nine pseudoscalar resonances in an $SU(3)$ symmetric situation (same masses for the pseudoscalars), eight resonances are degenerate and form an octet of scalar resonances and there is another one, with a different mass, which is an $SU(3)$ singlet. In particular, in the chiral limit ($m_\pi = m_K = 0$) Ref. [19] found that the octet pole position is around $500 - i\,350$ MeV and the singlet one is lighter at around $400 - i\,25$ MeV. Notice how different these pole positions are as compared to those in the actual physical situation, for which the Ref. [19] finds the poles that are given in Table 8.1. The strong dependence of the σ mass with the pion mass is confirmed recently by the lattice QCD calculation of Ref. [44]. In Table 8.1 we also give the coupling constants ξ_i to the different channels. These couplings are given by the calculation of the residues of the T matrix of PWAs at the pole position,

$$\xi_i \xi_j = \lim_{s \rightarrow s_R} (s - s_R) T_{ij}(s) , \qquad (8.26)$$

where the subscripts i and j indicate the coupled channels and s_R is the resonance pole.

It is also interesting to connect the formalism developed in Chaps. 6 and 7 with the so-called on-shell factorization of the vertices in a Bethe–Salpeter equation as

Table 8.1 Pole positions [MeV] and moduli of the couplings [GeV] for the lightest scalar resonances by applying the formalism described in this section from Ref. [19]. The couplings to the $\pi\pi$ and $\eta_8\eta_8$ channels have been multiplied by $\sqrt{2}$ to correct for the unitarity normalization

σ	$f_0(980)$	$a_0(980)$	κ
$\sqrt{s_\sigma} = 45 - i\,221$	$\sqrt{s_{f_0}} = 987 - i\,14$	$\sqrt{s_{a_0}} = 1053 - i\,24$	$\sqrt{s_\kappa} = 779 - i\,330$
$\lvert\xi_{\pi\pi}\rvert = 4.25$	$\lvert\xi_{K\bar{K}}\rvert = 3.63$	$\lvert\xi_{K\bar{K}}\rvert = 5.50$	$\lvert\xi_{K\pi}\rvert = 5.00$
$\left\lvert\frac{\xi_{K\bar{K}}}{\xi_{\pi\pi}}\right\rvert = 0.25$	$\left\lvert\frac{\xi_{\pi\pi}}{\xi_{K\bar{K}}}\right\rvert = 0.51$	$\left\lvert\frac{\xi_{\pi\eta_8}}{\xi_{K\bar{K}}}\right\rvert = 0.70$	$\left\lvert\frac{\xi_{K\eta_8}}{\xi_{K\pi}}\right\rvert = 0.62$
$\left\lvert\frac{\xi_{\eta_8\eta_8}}{\xi_{\pi\pi}}\right\rvert = 0.04$	$\left\lvert\frac{\xi_{\eta_8\eta_8}}{\xi_{K\bar{K}}}\right\rvert = 1.11$		

developed in Ref. [10].[4] In this reference the scalar lowest order ChPT amplitudes with $I = 0$ and 1 are unitarized by iterating them in a Bethe–Salpeter equation

$$T(k, p) = V(k, p) - i \int \frac{d^4q}{(2\pi)^4} \frac{V(k, q)T(q, p)}{[q^2 - m_1^2][(P - q)^2 - m_2^2]},$$ (8.27)

where we only give one four-momentum (let us say that of the first particle) for the initial (right) and final (left argument) states, because the other four-momenta can be obtained by the conservation of the total momentum P, which in the CM is $P = (\sqrt{s}, 0)$. In Eq. (8.27) $V(k, q)$ is the S-wave projected off-shell LO ChPT meson–meson scattering amplitude. The off-shell isoscalar scattering amplitude can be calculated straightforwardly from the LO ChPT Lagrangian [29], with the result [10]

$$V_{11}(s) = \frac{1}{9f^2}\left(9s + \frac{15}{2}m_\pi^2 - 3\sum_{i=1}^{4} p_i^2\right).$$ (8.28)

Notice that at this order it is just a purely S wave scattering amplitude. In the previous equation, p_i stands for the four-momentum of any pion and f is the bare pion decay constant (equal to f_π in the chiral limit). The whole on-shell factorization process in Ref. [10] is based on the observation that the off-shell part in the LO meson–meson scattering amplitudes considered, like the one in Eq. (8.28), is of the form $\sum_i c_i(p_i^2 - m_i^2)$, where the c_i are constants. Therefore, these off-shell terms cancel one or two of the propagators in the explicit loop function of the Bethe–Salpeter equation in Eq. (8.27), so that one ends with pure contact interactions without momentum flow. This type of diagrams only contributes to the renormalization of the bare parameters that appear already in the lowest order amplitudes [10, 45], and

---

[4]Our convention for the scattering amplitudes differs from that in Ref. [10] by a minus sign. There is also a reshuffling in the labeling of the states, so that we designate them with a label that increases as the associated threshold does.

whose values are rather constraint by phenomenology. These parameters can be taken at lowest order in the chiral expansion to be given by their physical values. In this way, Ref. [10] expresses the PWA $T(s)$ in the form,

$$T(s) = \left[\mathcal{N}(s)^{-1} + g(s)\right]^{-1} , \tag{8.29}$$

where $\mathcal{N}(s)$ is given now by the on-shell LO PWAs calculated in ChPT ($p_i^2 = m_i^2$). E.g., for the $I = L = 0$ meson–meson scattering one has

$$V_{11}(s) = \frac{1}{f_\pi^2}(s - \frac{m_\pi^2}{2}) , \tag{8.30}$$

$$V_{12}(s) = \frac{\sqrt{3}s}{4f_\pi^2} ,$$

$$V_{22}(s) = \frac{3s}{4f_\pi^2} ,$$

and $V_{21}(s) = V_{12}(s)$.

Chapter 9
$SU(3)$ Analysis of the Subtraction Constants in $g_i(s)$

In the $SU(3)$ limit the masses of all the hadrons belonging to a given $SU(3)$ multiplet have the same value. By direct product of monoparticle states of particles belonging to different $SU(3)$ multiplets we have multiparticle scattering states. In particular, for the two-body interaction process involving such particles we would have only to distinguish between the common masses of the particles in the $SU(3)$ representation involved. For instance, for the lightest pseudoscalar–pseudoscalar scattering we would have only one mass since all the particles belong to the same octet $SU(3)$ representation. Other two-body states of interest for our purposes is the one made by a baryon and one of the lightest pseudoscalars, all of them belonging to octet representations.

It is clear that because of the Wigner–Eckart theorem the matrix element of $SU(3)$ operators transforming within a given $SU(3)$ multiplet between states belonging to definite representations are independent of the hypercharge and third component of isospin, that characterize the different states and operators in a given irreducible representation [46]. The T matrix is an $SU(3)$ singlet and therefore the scattering matrix is diagonal in a basis of states with definite transformation properties under $SU(3)$. Denoting these states by $|R, \lambda\rangle$, with R corresponding to the $SU(3)$ irreducible representation and λ including the other quantum numbers needed to distinguish between states within R (e.g., third component of isospin and hypercharge). The momenta and spin indices are not indicated in the following since they do not play any active role in the next considerations. We then have for the T matrix,

$$\langle R', \lambda' | T | R, \lambda \rangle = T_R \delta_{RR'} \tag{9.1}$$

$$= \frac{1}{\mathcal{N}_R^{-1} + g_R(s)} \delta_{RR'}.$$

Here we have used the general parameterization of Eq. (7.2) with the unitarity function $g_R(s)$ containing the subtraction constant a_R.

© The Author(s), under exclusive licence to Springer Nature Switzerland AG 2019
J. A. Oller, *A Brief Introduction to Dispersion Relations*, SpringerBriefs in Physics,
https://doi.org/10.1007/978-3-030-13582-9_9

Let us further denote by $|i\rangle$ the physical states in the charged basis and the Clebsch–Gordan coefficients connecting both bases by $\langle i, R\lambda \rangle$. These real coefficients satisfy the orthogonality relations

$$\sum_i \langle i, R\lambda \rangle \langle i, R'\gamma \rangle = \delta_{RR'}\delta_{\lambda\gamma} , \tag{9.2}$$

$$\sum_{R,\lambda} \langle i, R\lambda \rangle \langle j, R\lambda \rangle = \delta_{ij} .$$

Notice that since $\langle i, R\lambda \rangle = \langle i|R, \lambda \rangle$ and is real, then it also follows that $\langle i, R\lambda \rangle = \langle R, \lambda|i \rangle$.

In the physical basis the T-matrix elements $T_{ij}(s)$ also obeys Eq. (7.2), with $\mathcal{N}(s)$ calculated in the charged basis and the functions $g_i(s)$ involving the subtraction constants a_i. Let us show that all the a_R and a_i have the same value in the $SU(3)$ limit, as derived in Ref. [47]. For that we proceed with the change of basis of a singlet $SU(3)$ matrix A, from the physical basis to the $SU(3)$ one. Then,

$$\sum_{ij} \langle i, R\lambda \rangle A_{ij} \langle j, R'\gamma \rangle = A_R \delta_{RR'}\delta_{\lambda\gamma} . \tag{9.3}$$

For instance, this is case for the QCD Hamiltonian, and therefore for the T matrix, as well as for the unitarity loop function $g_i(s)$, Eq. (8.5). The latter function, contrary to the T matrix, is also diagonal in the physical basis. This is a key distinctive feature that allows us to perform the following manipulations. By inverting Eq. (9.3) with $g(s)$ instead of A, we have that

$$g_i(s)\delta_{ij} = \sum_{R,\lambda} \langle i, R\lambda \rangle g_R(s) \langle j, R\lambda \rangle . \tag{9.4}$$

Next, we multiply by $\langle j, R'\gamma \rangle$ and sum over j,

$$g_i(s)\langle i, R'\gamma \rangle = \sum_{R,\lambda}\sum_j \langle i, R\lambda \rangle g_R(s) \underbrace{\langle j, R\lambda \rangle \langle j, R'\gamma \rangle}_{\delta_{RR'}\delta_{\lambda\gamma}} = \langle i, R'\gamma \rangle g_{R'}(s) . \tag{9.5}$$

From this relation it is sufficient to take physical states with components in different irreducible representation to conclude that for any state $|i\rangle$ in the charged basis and for any irreducible $SU(3)$ representation R involved in the decomposition of the charged basis in $SU(3)$ multiplets, one has

$$g_i(s) = g_R(s) = g(s) . \tag{9.6}$$

As a result it follows the equality in the $SU(3)$ limit of all the subtraction constants for the two-particles states $|AB\rangle$, with A and B belonging to the irreducible $SU(3)$ representations R_A and R_B, in order.

This result is manifestly evident when every subtraction constant $a_i(\mu)$ is given by its natural value, Eq. (8.8), because then the masses m_1 and m_2 (as well as the three-momentum cut-off Λ) are common to all the two-particle states in the $SU(3)$ limit.

Chapter 10
Perturbative Introduction
of Crossed-Channel Cut Singularities

Let us now consider the perturbative treatment of the crossed-channel cuts, that for simplicity are denoted generically as LHC, when unitarizing PWAs obtained from some EFT. We present four methods; the first one is based in the use of $\mathcal{N}(s)$ as introduced in Eq. (7.1). From this method we also derive another approach that is referred in the literature as the Inverse Amplitude Method (IAM). Other two approaches arise from the use of the N/D method where $\Delta(p^2)$ is calculated perturbative in the considered EFT, such that either the N/D IE is solved fully or in its first iterated form. This line of handling perturbatively the LHC contributions is discussed in Chaps. 11 and 12, in order.

Let us suppose that $T(s)$ is given by Eq. (7.2) in terms of $\mathcal{N}(s)$ and $g(s)$. In view of Eq. (7.10) a convenient choice for the subtraction constant in the unitarity loop function would be such that $g(s)$ is zero at some point along the near-threshold LHC. In this way, we might dismiss the dependence of $\mathcal{N}(s)$ along the physical region ($s \geq s_{\text{th}}$) on the iterated LHC contributions [multiplied by $g(s)$], at least for not too high s. For example, if one imposes that $g(0) = 0$ then, by taking the subtraction point at $s = 0$, the subtraction constant would be simply zero. Had we imposed that $g(s_0) = 0$, with s_0 along the LHC, then we would change the subtraction point to s_0, so that again $a(s_0) = 0$. In this case $g(s)$ would read

$$g(s) = -\frac{s - s_0}{\pi} \int_{s_{\text{th}}}^{\infty} \frac{\rho(s')ds'}{(s' - s_0)(s' - s)} \tag{10.1}$$
$$= -\frac{s}{\pi} \int_{s_{\text{th}}}^{\infty} \frac{\rho(s')ds'}{s'(s' - s)} + \frac{s_0}{\pi} \int_{s_{\text{th}}}^{\infty} \frac{\rho(s')ds'}{s'(s' - s_0)} \, .$$

This choice for the subtraction constant in order to weaken the influence of the iterated LHC for low s might be of relevance if one wished to provide a perturbative solution of Eq. (7.10) for $\mathcal{N}(s)$. Indeed, from Eq. (7.2) we have the geometric series

© The Author(s), under exclusive licence to Springer Nature Switzerland AG 2019
J. A. Oller, *A Brief Introduction to Dispersion Relations*, SpringerBriefs in Physics,
https://doi.org/10.1007/978-3-030-13582-9_10

$$T = \mathcal{N} - \mathcal{N}g\mathcal{N} - \mathcal{N}g\mathcal{N}g\mathcal{N} + \ldots \tag{10.2}$$

This expansion in powers of $\mathcal{N}g$ could be matched with a perturbative loop expansion of T and, in this way, $\mathcal{N}$ could be determined order by order [8, 48].

Let us give an explicit example based on ChPT applied in Ref. [48] to the massless $W_L W_L$ scattering (the subscript L stands for the longitudinal component of the W boson) by applying the equivalence theorem [49]. In this EFT the momentum expansion implies a loop expansion, so that the chiral dimension of a perturbative Feynman graph with L loops is $D = 2L + 2 + \sum_d N_d(d-2)$, with d the chiral dimension of a given monomial in the ChPT Lagrangian and N_d is the number of such vertices with dimension d [50]. The isoscalar scalar $W_L W_L$ scattering amplitude up to NLO or $\mathcal{O}(p^4)$ in ChPT is [48, 51]

$$T_2(s) = \frac{s}{v^2}, \tag{10.3}$$

$$T_4(s) = \frac{3s^2}{2v^2(m_H^2 - s)} + \frac{m_H^4}{v^2 s}\left[\log\left(1 + \frac{s}{m_H^2}\right) - \frac{s}{m_H^2} + \frac{s^2}{2m_H^4}\right] \tag{10.4}$$
$$- \frac{s^2}{1728\pi^2 v^4}\left[1673 - 297\sqrt{3}\,\pi + 108\log\frac{-s}{m_H^2} + 42\log\frac{s}{m_H^2}\right],$$

where $v = (\sqrt{2}G_F)^{-1/2} \simeq 1/4$ TeV is the analogous to f_π for the pion case, with G_F the Fermi coupling constant. The expression for $T_4(s)$ in Eq. (10.4) contains the exchange of a Standard Model Higgs boson of mass m_H. If we denoted by b the combination

$$\frac{11v^2}{6m_H^2} - \frac{1673 - 297\pi\sqrt{3}}{1728\pi^2} \to b, \tag{10.5}$$

then the expression in Eq. (10.4) becomes more general and it does not necessarily correspond to the exchange of a Standard Model Higgs boson, but to a general scenario of another underlying fundamental theory. The amplitude $T_4(s)$, up to $\mathcal{O}(p^4)$, becomes then

$$T_4(s) = b\frac{s^2}{v^4} - \frac{s^2}{1728\pi^2 v^4}\left[108\log\frac{-s}{m_H^2} + 42\log\frac{s}{m_H^2}\right]. \tag{10.6}$$

At NLO all of these alternative theories would give rise to $T_4(s)$ as written above in terms of v and b, although with the latter having different values. The scale m_H^2 is introduced above to refer to a high-energy scale in which bare resonance could appear. In the previous equation the first logarithm gives rise to the RHC and the last to the LHC.

In order to proceed with the unitarization of $T_2(s) + T_4(s)$ we employ the non-perturbative expression

$$T(s) = \frac{\mathcal{N}(s)}{1 + g(s)\mathcal{N}(s)} \,, \tag{10.7}$$

equivalent to Eq. (7.10). Next, we proceed with the chiral expansion of $\mathcal{N}(s)$ up to NLO as

$$\mathcal{N}(s) = \mathcal{N}_2(s) + \mathcal{N}_4(s) + \mathcal{O}(p^6) \tag{10.8}$$

with the subscript indicating the chiral order. Then, we match the chiral expansion of $T(s)$ in Eq. (10.7) by counting the loop function $g(s)$ as $\mathcal{O}(p^0)$, as it is clear from its loop expression in Eq. (8.5). This function in the present massless case reads

$$g(s) = \frac{1}{16\pi^2}\left(a + \log\frac{-s}{m_H^2}\right) . \tag{10.9}$$

Therefore we have,

$$\mathcal{N}_2(s) = T_2(s) \,, \tag{10.10}$$
$$\mathcal{N}_4(s) = T_4(s) + T_2(s)^2 g(s) \tag{10.11}$$
$$= \frac{s^2}{288\pi^2 v^4}\left(18(a + 16b\pi^2) - 7\log\frac{s}{m_H^2} .\right) .$$

In this way we can employ Eq. (10.7) to calculate $T(s)$ by using $\mathcal{N} = \mathcal{N}_2 + \mathcal{N}_4$, the latter ones determined in Eqs. (10.10) and (10.11). In the limit in which $m_H \gg 4\pi v$, while keeping $|a|$ and $|b|$ of $\mathcal{O}(1)$ [b is around 0.1 for a heavy Standard Model Higgs boson of mass 1 TeV, cf. Eq. (10.5)], an isoscalar scalar resonance with vanishing mass and width is dynamically generated [48, 52, 53]. In the opposite limit, $m_H \ll 4\pi v$, we consider again the perturbative expression for $T_2(s) + T_4(s)$ given in Eqs. (10.3) and (10.4) corresponding to the exchange of a standard model Higgs. Further, we neglect the non-logarithmic terms divided by $4\pi v$ in comparison with those divided by the much smaller m_H. Additionally, near the bare pole, $s \simeq m_H^2$, the direct exchange of the resonance dominates over other contributions and, after these simplifications, we have now for $\mathcal{N}(s)$

$$\mathcal{N}(s) \approx \frac{3s^2}{2v^2(m_H^2 - s)} . \tag{10.12}$$

When inserted in Eq. (10.7) for calculating the unitarized the PWA $T(s)$ we obtain

$$T(s) \approx \frac{3m_H^2/(2v^2)}{m_H - \sqrt{s} - i\frac{3m_H^3}{64\pi v^2}} . \tag{10.13}$$

In this form, we end with a Breit–Wigner parameterization for the Higgs exchanged, with mass m_H and width

$$\Gamma_H = \frac{3m_H^3}{32\pi v^2} , \tag{10.14}$$

which coincides with the QFT expression for the width of the Higgs boson from the electroweak Lagrangian. This width is much smaller than m_H for $m_H \ll 4\pi v$.

In the literature there have been many other studies in which the LHC is included perturbatively and that could be understood by employing the basic point in the expansion given in Eq. (10.2).

Let us consider first the Inverse Amplitude Method, on which we briefly report. We come back again to Eq. (7.2) an express $\mathcal{N} = \mathcal{N}_2 + \mathcal{N}_4 + \mathcal{O}(p^6)$, with the former given by Eq. (10.10) and the latter by the first line Eq. (10.11), after matching with $T = T_2 + T_4 + \mathcal{O}(p^6)$ as explained above. Then,

$$T(s) = \left(\left[T_2(s) + T_4(s) + T_2(s)g(s)T_2(s) + \mathcal{O}(p^6) \right]^{-1} + g(s) \right)^{-1} . \tag{10.15}$$

We perform next the chiral expansion of the inverse matrix between the square brackets

$$T(s) = \left(T_2(s)^{-1} \left[I + T_4(s)T_2(s)^{-1} + T_2(s)g(s) + \mathcal{O}(p^4) \right]^{-1} + g(s) \right)^{-1} \tag{10.16}$$

$$= \left(T_2(s)^{-1} \left[I - T_4(s)T_2(s)^{-1} - T_2(s)g(s) + \mathcal{O}(p^4) \right] + g(s) \right)^{-1} \tag{10.17}$$

$$= \left[I - T_4(s)T_2(s)^{-1} + \mathcal{O}(p^4) \right]^{-1} T_2(s)$$

$$= T_2(s) \left[T_2 - T_4 + \mathcal{O}(p^6) \right]^{-1} T_2(s) .$$

The last expression corresponds to the NLO IAM [32, 54–58]

$$T(s) = T_2(s) \left[T_2(s) - T_4(s) \right]^{-1} T_2(s) . \tag{10.18}$$

Despite it is based on a perturbative solution of Eq. (7.10), the IAM result is independent of the subtraction constant in $g(s)$. That this should be the case is clear if one considers that the IAM can also be recast as the expansion of the inverse of the PWA, $T(s)^{-1} = (T_2 + T_4)^{-1} = T_2^{-1}(T_2 - T_4 + \mathcal{O}(p^6))T_2^{-1}$, and then taking the inverse of this expansion.

There is an alternative derivation of the uncoupled IAM based on a DR for the inverse PWA $T^{-1}(s)$ [54, 55, 57]. Instead of taking directly $1/T(s)$ one consider the auxiliary function $G(s) = T_2(s)^2/T(s)$, whose imaginary part is, cf. Eq. (2.51),

$$\Im G = -T_2(s)^2 \rho(s) . \tag{10.19}$$

We write down a three-times subtracted DR for $G(s)$ by applying the Sugawara–Kanazawa theorem discussed in Chap. 4, because $T_2(s)^2$ at most diverges like s^2 and $T(s) \to$ constant for $s \to +\infty \pm i\varepsilon$ because of unitarity. It then follows that

$$G(s) = G(0) + G'(0)s + \frac{1}{2}G''(0)s^2 - \frac{s^3}{\pi}\int_{s_{th}}^{\infty} ds' \frac{\rho(s')T_2(s')^2}{(s')^3(s'-s)} - LC(G) + PC(s) ,$$

$$(10.20)$$

where $-LC(G)$ refers of the crossed-channel contributions in $G(s)$ and the pole contributions $PC(s)$ arise from possible zeroes of $T(s)$. We neglect this last contribution because the zeroes in the denominator are largely canceled by $T_2(s)^2$ when forming $G(s)$. There could be some slight mismatch between the zeroes of $T_2(s)$ and those of $T(s)$ which might give rise to a pathological behavior in narrow energy regions [32, 58, 59]. A modified version of the IAM formula was derived in Ref. [59] to cure this deficiency. Notice also that the expression for $T = 1/([N_2 + N_4]^{-1} + g)$, without the expansion of $[N_2 + N_4]^{-1}$, has no this pathology.

The subtraction constants $G(0)$, $G'(0)$ and $G''(0)$ are fixed by matching with the ChPT expansion of $G(s)$ up NLO,

$$G(s) = \frac{T_2(s)^2}{T_2(s) + T_4(s) + \mathcal{O}(p^6)} = T_2(s) - T_4(s) + \mathcal{O}(p^6) . \qquad (10.21)$$

Therefore, by neglecting higher orders we can identify $G(0) = T_2(0) - T_4(0)$, $G'(0) = T_2'(0) - T_4'(0)$, and $G''(0) = T_2''(0) - T_4''(0)$. By the same token, $LC(G)$ is approximated from the crossed-channel cut contribution of $T_4(s)$, $LC(T_4)$. Furthermore, the dispersive integral in Eq. (10.20) is minus the one for the RHC contribution in $T_4(s)$, whose imaginary part along the RHC is $\Im T_4(s) = T_2(s)^2\rho(s)$, as required by perturbative unitarity. Thus, Eq. (10.20) becomes $G(s) = T_2(s) - T_4(s)$ and then $T(s)$ is given by Eq. (10.18). The IAM has also been extended to two-loop ChPT amplitudes in Ref. [60]. This method has been applied to meson–meson scattering [32, 57, 58, 61], quark-mass dependence of masses and decay constants [62], $W_L W_L$ scattering [52, 53], πN scattering [63], etc., among many other references.

Chapter 11
The N/D Method with Perturbative $\Delta(p^2)$

Here we take the point of view of calculating perturbatively in an EFT the discontinuity of a PWA along the LHC, that we denoted above as $\Delta(p^2)$. Once $\Delta(p^2)$ is approximated in this way one can then solve the IE that follows from the N/D method in order to calculate $D(s)$ along the LHC, and then the full $T(s)$. We mostly follow here the presentation in Ref. [64], where the N/D method is applied to the study of the NN interactions in nonrelativistic scattering theory, employing ChPT Lagrangians with 0, 1 and 2 conserved baryon number. The PWAs in this case are characterized as in Chap. 2, by the total angular momentum J, total spin S and orbital angular momentum ℓ. There could be mixing in some triplet states $(S = 1)$ between the partial waves with $\ell = J \pm 1$.

The NN PWAs are analytic functions of p^2 with a RHC and a LHC. The onset of the LHC is due to one-pion exchange in the t- and u-channels when the pion-exchange propagator is on-shell, $t = m_\pi^2$ or $u = m_\pi^2$, respectively. Therefore, it extends for

$$p^2 \leq -\frac{m_\pi^2}{4} \equiv L \ . \tag{11.1}$$

Within nonrelativistic kinematics $p^2 = mE$, with $E = \sqrt{s} - 2$, the NN kinetic energy.

We present here the N/D method with some differences with respect to Chap. 6 because the LHC contribution is not neglected and then, a different way to guarantee the required threshold behavior of a PWA (which should vanish as $p^{2\ell}$ for $p \to 0$) is convenient. Our basic equations are

$$T(p^2) = \frac{N(p^2)}{D(p^2)} \ , \tag{11.2}$$

$$\Im D(p^2) = -\rho(p^2)N(p^2) \ , \ p^2 \geq 0 \ ,$$

$$\Im N(p^2) = D(p^2)\Delta(p^2) \ , \ p^2 \leq L \ ,$$

© The Author(s), under exclusive licence to Springer Nature Switzerland AG 2019
J. A. Oller, *A Brief Introduction to Dispersion Relations*, SpringerBriefs in Physics,
https://doi.org/10.1007/978-3-030-13582-9_11

being zero the imaginary parts of N and D for $p^2 > L$ and $p^2 < 0$, in order. We also take the standard nonrelativistic normalization for a PWA

$$\rho(p^2) = \frac{m\sqrt{p^2}}{4\pi} , \qquad (11.3)$$

the same as already considered in the nonrelativistic partial-wave expansion of the LS equation, compared, e.g., with Eq. 2.82 for $\mu = m/2$. In this equation the prefactor $\mu k/2\pi$ then becomes $\rho(k^2 + i\varepsilon)$.

The N/D equations with m subtractions in N and n subtractions in D are referred as ND_{mn}. They can be written as,

$$D(p^2) = 1 + \sum_{i=1}^{n-1} \delta_i (p^2 - p_0^2)^i - \frac{(p^2 - p_0^2)^n}{\pi} \int_0^\infty \frac{\rho(q^2) N(q^2) dq^2}{(q^2 - p_0^2)^n (q^2 - p^2)} , \quad (11.4)$$

$$N(p^2) = \sum_{i=0}^{m-1} \nu_i (p^2 - p_0^2)^i + \frac{(p^2 - p_0^2)^m}{\pi} \int_{-\infty}^L \frac{\Delta(q^2) D(q^2) dq^2}{(q^2 - p_0^2)^m (q^2 - q^2)} . \qquad (11.5)$$

Where we have taken advantage that the ratio of N and D is unchanged if these functions are renormalized by the same factor to fix that $D(p_0^2) = 1$. Certainly, one could introduce several subtraction points and not always the same one, p_0^2, as written above. The standard N/D equations result after substituting the expression for N in that of D, which then reads

$$D(p^2) = 1 + \sum_{i=1}^{n-1} \delta_i (p_i^2 - p_0^2)^i - \sum_{i=0}^{m-1} \nu_i \frac{(p^2 - p_0^2)^n}{\pi} \int_0^\infty \frac{\rho(q^2) dq^2}{(q^2 - p^2)(q^2 - p_0^2)^{n-i}}$$

$$(11.6)$$

$$+ \frac{(p^2 - p_0^2)^n}{\pi^2} \int_{-\infty}^L \frac{\Delta(q^2) D(q^2)}{(q^2 - p_0^2)^m} \int_0^\infty \frac{\rho(k^2) dk^2}{(k^2 - p^2)(k^2 - q^2)(k^2 - p_0^2)^{n-m}} .$$

We impose in the following that $m \leq n$ because otherwise the last integration on the RHC in the previous equation is divergent.

The previous DR provides us with a linear IE to calculate $D(p^2)$ for $p^2 < L$, which we solve numerically by discretizing the integration and inverting the resulting matrix. Once $D(p^2)$ is known along the LHC then we can calculate the functions $D(p^2)$ and $N(p^2)$ in the whole complex p^2 plane by employing Eq. (11.6) and Eq. (11.5), respectively. As a result of this knowledge, we can then determine the PWA $T(p^2) = N(p^2)/D(p^2)$ as a function of the complex p^2 variable.

All the integrals along the RHC in Eq. (11.6) can be done algebraically. They can be written out of the unitarity loop function $g(p^2, k^2)$, which is defined as

$$g(p^2, k^2) = \frac{m}{4\pi^2} \int_0^\infty \frac{q dq^2}{(q^2 - p^2)(q^2 - k^2)} = \frac{im/4\pi}{\sqrt{p^2} + \sqrt{k^2}} . \qquad (11.7)$$

For p^2 and k^2 along the LHC we should use the previous expression replacing $p^2 \to p^2 + i\varepsilon$ and analogously for k^2. By differentiating $g(p^2, k^2)$ with respect to k^2 we can calculate the RHC integrals with more subtractions involved. Namely,

$$\frac{\partial^j g(p^2, k^2)}{\partial k^{2j}} = \frac{j!}{\pi} \int_0^\infty \frac{\rho(q^2)dq^2}{(q^2 - p^2)(q^2 - k^2)^{j+1}} . \tag{11.8}$$

In Refs. [64, 65] and [66] $\Delta(p^2)$ was calculated at the chiral orders p, p^2 and p^3, respectively, from the chiral perturbative calculation of the NN scattering amplitude in Ref. [67]. A good reproduction of the NN phase shifts is achieved already at $\mathcal{O}(p^3)$.

A way to impose by construction that a PWA with orbital angular momentum ℓ vanishes in the limit $p^2 \to 0$ as $p^{2\ell}$ is to add a higher order CDD pole in the D function as $\gamma_\ell/p^{2\ell}$ in Eq. (11.6). This is the minimum input to keep the required behavior at threshold given the fact that $\Delta(p^2)$ is only calculated perturbatively and we are later using it to evaluate a non-perturbative amplitude [7, 65]. One could also include more terms, and write $\sum\limits_{i=1}^{\ell} \dfrac{\gamma_i}{p^{2i}} + 1 + \sum\limits_{i=1}^{n-1} \delta_i p^{2i}$ instead of just $1 + \sum\limits_{i=1}^{n-1} \delta_i p^{2i}$ in Eq. (11.6). The number of subtraction constants would be limited depending on the problem, e.g., by requiring to reproduce a few low-energy parameters associated with an ERE.

An interesting point from the N/D method, as found out in Ref. [66], is that it allows a very precise determination of the ERE shape parameter up to very high orders. The improvement stems from the fact that this method provides us with the PWA in the complex p^2 plane, so that one can apply the Cauchy theorem to calculate derivatives with high precision. Notice that one can express them in terms of an integral and use in its numerical evaluation large numbers of points when discretizing the integral. More precisely, given an elastic PWA $t(s)$ with orbital angular momentum ℓ and normalized as

$$t(p^2) = \frac{1}{p \cot \delta_\ell - ip} , \tag{11.9}$$

$$t(p^2) = \frac{m T(p^2)}{4\pi} = \frac{m}{4\pi} \frac{N(p^2)}{D(p^2)} ,$$

the ERE is given by

$$p^{2\ell+1} \cot \delta_\ell = -\frac{1}{a} + \frac{1}{2} r p^2 + \sum_{i=2}^\infty v_i p^{2i} , \tag{11.10}$$

where this expansion is done at threshold and it is convergent until the nearest singularity of T^{-1} [once the threshold branch point is discounted because it is removed from $p \cot \delta_\ell$ (also called the the K matrix) in Eq. (11.9)]. In NN scattering this is the starting of the one-pion exchange threshold, but in other physical systems the radius of convergence could be settled by a near-threshold zero of $t(p^2)$ [in other terms because of a CDD pole].

Following Ref. [66] we define the function

$$H(p^2) = \frac{4\pi}{m} \frac{D(p^2)}{N(p^2)} p^{2\ell} + i p^{2\ell} \sqrt{p^2} = p^{2\ell} \sqrt{p^2} \cot \delta , \qquad (11.11)$$

with $\sqrt{p^2}$ defined in the first Riemann sheet. The derivatives of $H(p^2)$ at the origin provides us with the different terms in the ERE. Denoting by $H^{(n)}(0)$ its n_{th} derivative, we have from the Cauchy integral formula

$$H^{(n)}(0) = \frac{n!}{2\pi i} \oint_C \frac{H(z)dz}{z^{n+1}} , \qquad (11.12)$$

where C can be any close contour inside the radius of convergence of the ERE expansion, $p^2 < L$. In practice we always take a circle of radius $r < L$ and, for testing numerically the outcome, we vary r within this limit and check the stability of the numerical results. As a result

$$a^{-1} = -\frac{1}{2\pi i} \oint_C \frac{H(z)dz}{z} , \qquad (11.13)$$

$$r = \frac{1}{\pi i} \oint_C \frac{H(z)dz}{z^2} ,$$

$$v_i = \frac{1}{2\pi i} \oint_C \frac{H(z)dz}{z^{i+1}} .$$

For coupled PWAs one could proceed analogously but using the eigenvalues of the S matrix. Making use of this method Ref. [66] is able to calculate a shape parameter of an order as high as 10, v_{10}, with a numerical precision around 1%.

A more formal point established in Ref. [64], by making use of the Fredholm theory for linear IEs [18], is to elaborate a compelling argument to conclude that if the function $\Delta(p^2)$ behaves asymptotically for $p^2 \to -\infty$ as $p^{2\gamma}$ with $\gamma < -1/2$, then any ND$_{mn}$ IE has solution. The perturbative $\Delta(p^2)$ calculated in a low-energy EFT typically scales like a power of p^2 (modulo logarithmic factors), and indeed it usually rises with the higher order of the calculation. This is the case for $\Delta(p^2)$ calculated in ChPT for NN. At LO it vanishes at least as $1/A$, at NLO it diverges at most like A and at next-to-next-to-leading order as $A^{3/2}$.

Let us outline the main steps in the derivation given in Ref. [64]. We take for $\Delta(p^2)$ its asymptotic behavior, $\Delta(p^2) \to \lambda(-p^2)^\gamma$, and start with the once-subtracted DR form (ND$_{11}$) for the IE of the function $D(p^2)$ along the LHC, $p^2 < L$,

$$D(p^2) = 1 + \nu_1 \frac{m\sqrt{-p^2}}{4\pi} + \frac{mp^2}{4\pi^2} \int_{-\infty}^{L} \frac{\Delta(q^2)D(q^2)dq^2}{q^2(\sqrt{-q^2} + \sqrt{-p^2})} . \qquad (11.14)$$

At this stage we replace $\Delta(p^2)$ by its asymptotic behavior and introduce the new variables $x = L/q^2$ and $y = L/p^2$. The D function is also multiplied by $y^{-\gamma/2}$, $\tilde{D} = y^{-\gamma/2}D$, to obtain an IE with a symmetric kernel in the two variables x and y. In this way, we have from Eq. (11.14)

$$\tilde{D}(y) = y^{-\gamma/2} + y^{-\frac{\gamma+1}{2}}\nu_1 \frac{m(-L)^{\frac{1}{2}}}{4\pi} + \frac{\lambda m}{4\pi^2}(-L)^{\gamma+\frac{1}{2}} \int_0^1 \frac{\tilde{D}(x)dx}{(xy)^{\frac{\gamma+1}{2}}\left(x^{\frac{1}{2}} + y^{\frac{1}{2}}\right)} .$$

$$(11.15)$$

The symmetric kernel and the independent term for this IE are, in order,

$$K(y, x) = \frac{1}{(xy)^{\frac{\gamma+1}{2}}\left(x^{\frac{1}{2}} + y^{\frac{1}{2}}\right)} , \qquad (11.16)$$

$$f(y) = y^{-\frac{\gamma}{2}} + y^{-\frac{\gamma+1}{2}}\nu_1 \frac{m(-L)^{\frac{1}{2}}}{4\pi} .$$

Both $K(x, y,)$ and $f(y)$ are quadratically integrable[1] for $\gamma < -1/2$ and then, because of the Fredholm theorem, we conclude that the IE of Eq. (11.15) has a unique solution for $\gamma < -1/2$.[2] This result can be easily extended to any IE of the type ND_{mn}, following again similar steps as explained in Ref. [64] for the case ND_{nn} (since for $m = n$ the largest powers of q^2 happen in the integrals along the RHC, the IEs with $m < n$ are more convergent). The argument used in Ref. [64] goes as follows. If we increase by one subtraction both N and D then we have a one more power of p^2 multiplying the integral in Eq. (11.14), one more factor of k^2 when inserting N in the expression for D and another extra factor of $1/k^2$ from the RHC integral of D, cf. Eqs. (11.4) and (11.5). The latter two factors cancel and the former gives rise to an extra factor of L/y. This is compensated when multiplying D by y to end with the same symmetric kernel $K(y, x)$ of Eq. (11.16). Regarding the independent term, it does not become more singular either because the extra power of $1/y$ from the last subtraction is compensated when the IE is multiplied by one extra factor of

---

[1] By this we mean that

$$\int_0^1 dy \int_0^1 dx \, K(x, y)^2 < \infty , \qquad (11.17)$$

$$\int_0^1 dy \, f(y)^2 < \infty .$$

[2] The only exception might be if the factor $\frac{\lambda m}{4\pi^2}(-L)^{\gamma+\frac{1}{2}}$ multiplying the integral in Eq. (11.15) coincided with an eigenvalue of the kernel $K(y, x)$. Nonetheless, since λ is continuous and the eigenvalues of a kernel are discrete we could employ a smooth continuation to find the solution in the a priori unlikely case of such coincidence.

y, so as to end with the IE with a symmetric kernel. This process can then be iterated up to the necessary number of subtractions and both the kernel and the independent term remain square integrable functions as they were in the ND_{11} case.

Chapter 12
The First-Iterated N/D Solution with Perturbative $\Delta(p^2)$

Given a discontinuity of a PWA along the LHC, $\Delta(p^2)$, the first-iterated N/D solution is obtained from the N/D DRs, cf. Eqs. (11.4) and (11.5), by iterating once along the RHC. This implies to substituting $D(q^2) \to 1$ in the equation for $N(p^2)$, Eq. (11.5). As a result the DR for $N(p^2)$ corresponds to the perturbative amplitude $V(p^2)$ with only LHC, which can be calculated by performing the corresponding integration, whose integrand is known now. Furthermore, the most common situation is that $V(p^2)$ is an input function already calculated from some theory. The integrand for $D(p^2)$ in Eq. (11.4) is also known. Then, it turns out that the first-iterated solution of the N/D method is

$$V(p^2) = \sum_{i=0}^{m-1} \nu_i (p^2 - p_0^2)^i + \frac{(p^2 - p_0^2)^m}{\pi} \int_{-\infty}^{L} \frac{\Delta(q^2) dq^2}{(q^2 - p_0^2)^m (q^2 - q^2)} , \qquad (12.1)$$

$$D(p^2) = 1 + \sum_{i=1}^{n-1} \delta_i (p^2 - p_0^2)^i - \frac{(p^2 - p_0^2)^n}{\pi} \int_{0}^{\infty} \frac{\rho(q^2) V(q^2) dq^2}{(q^2 - p_0^2)^n (q^2 - p^2)} . \qquad (12.2)$$

The applications for which this method is easier to implement are those for which $\Delta(p^2)$ stems from some tree-level amplitudes. Then, one can directly identify $V(p^2)$ with the resulting PWA stemming from these tree-level amplitudes.[1] The only input a priori unknown are the subtraction constants in $D(p^2)$. The minimum number of them is required by the diverging degree for $p^2 \to \infty$ of $V(p^2)$, so that if it diverges like $p^{2(m-1)}$ then m subtractions are needed, as written in Eq. (12.2). One could also explore adding more subtractions, depending on the requirements of the problem.

This scheme of work has been used recently to study $\rho\rho$ scattering in Ref. [9]. In the rest of this section we come back to use s instead of p^2 as argument of the PWAs, since the only DR involved is the one for $D(s)$ along the RHC, and the crossed-channel cuts are absorbed in the algebraic expression for $V(s)$. Thus, we then do not perform any explicit DRs on the crossed-channel cuts. Furthermore, the circular

---

[1] If there where some s-channel exchange of a bare particle this would give rise to a pole in $V(s)$ that could be straightforwardly accommodated by a adding a pole in the DR for $V(p^2)$, cf. Eq. (4.4).

© The Author(s), under exclusive licence to Springer Nature Switzerland AG 2019
J. A. Oller, *A Brief Introduction to Dispersion Relations*, SpringerBriefs in Physics,
https://doi.org/10.1007/978-3-030-13582-9_12

cuts are absent because this is a scattering problem with all the masses equal in the isospin limit, the one that we assume as valid in good approximation.

In Ref. [9] the dynamical input for the $\rho\rho$ interactions is given by the tree-level amplitudes obtained from the gauge-boson part of the hidden-gauge Lagrangian [36, 37]

$$\mathcal{L}' = -\frac{1}{4}\text{Tr}\,F_{\mu\nu}F^{\mu\nu} , \qquad (12.3)$$

$$F_{\mu\nu} = \partial_\mu V_\nu - \partial_\nu V_\mu - ig[V_\mu, V_\nu] ,$$

$$V_\mu = \begin{pmatrix} \frac{1}{\sqrt{2}}\rho^0 & \rho^+ \\ \rho^- & -\frac{1}{\sqrt{2}}\rho^0 \end{pmatrix} . \qquad (12.4)$$

The resulting vertices either involve three ρs or four of them. For $\rho\rho$ scattering the later is a contact interaction while from the three-ρs vertices one has the s-, t- and u-channel exchanges of one ρ. These latter contributions give rise to the LHC when projected in PWAs. This is the source of the perturbative $\Delta(s)$ in the present example, and the sum of all these tree-level diagrams (projected in the appropriate PWA) is $V(s)$ for a given reaction. The unitarization of $\rho\rho$ tree-level amplitudes obtained from the hidden-gauge chiral symmetry Lagrangian was pioneered by Ref. [68]. However, in this reference all the input tree-level interactions were reduced to contact ones without LHCs [that we denote by $\widetilde{V}(s)$], by taking the extreme nonrelativistic limit in the propagators of the ρ mesons. Nonetheless, this approach is consistent near the $\rho\rho$ threshold s_{th} and its unitarization is given by the Eq. (10.7) with $\mathcal{N}(s) = \widetilde{V}(s)$,

$$\widetilde{T}(s) = \frac{\widetilde{V}(s)}{1 + \widetilde{V}(s)g(s)} . \qquad (12.5)$$

Notice that in the simplifying approximation of Ref. [68] the LHC discontinuity is zero, $\Delta(s) = 0$, and then from Eq. (7.10) the function $\mathcal{N}(s)$ is just a polynomial. Since in the threshold region $V(s) \to \widetilde{V}(s)$ we also consider for subsequent use the analogous expression to Eq. (12.5) but replacing $\widetilde{V}(s)$ by $V(s)$,

$$\widetilde{A}(s) = \frac{V(s)}{1 + V(s)g(s)} . \qquad (12.6)$$

Now, the Eq. (12.6) is used in the near-threshold region to pin down the subtraction constants in Eq. (12.2). Three subtractions are taken because the amplitudes $V(s)$ diverge at most like s^2 for $s \to \infty$ [9]. We also rewrite Eq. (12.2) so that the subtraction polynomial is written in powers of $s - s_{\text{th}}$, because the matching with Eq. (12.6) is made around s_{th}. Namely, the resulting expression for $D(s)$ that we use is

$$D(s) = \gamma_0 + \gamma_1(s - s_{th}) + \frac{1}{2}\gamma_2(s - s_{th})^2 - \frac{(s - s_{th})s^2}{\pi}\int_{s_{th}}^{\infty}\frac{\rho(s')V(s')ds'}{(s' - s_{th})(s')^2(s' - s)} .$$

(12.7)

Its matching with the denominator of Eq. (12.6) gives rise to the equation

$$\gamma_0 + \gamma_1(s - \text{th}) + \frac{1}{2}\gamma_2(s - s_{th})^2 = 1 + V(s)g(s)$$

(12.8)

$$+ \frac{(s - s_{th})s^2}{\pi}\int_{s_{th}}^{\infty}\frac{\rho(s')V(s')ds'}{(s' - s_{th})(s')^2(s' - s)} .$$

In the following we denote by $\omega(s)$ the rhs of the previous equation and then we are left with the following expressions for the subtraction constants γ_i,

$$\gamma_0 = 1 + V(s_{th})g(s_{th}) ,$$

(12.9)

$$\gamma_1 = \omega'(s_{th}) ,$$

$$\gamma_2 = \omega''(s_{th}) ,$$

where we indicate by a prime the derivative of $\omega(s)$ with respect to s. The subtraction constant in $g(s)$ is taken according to its natural value, cf. Eq. (8.8). Making use of this unitarization approach Ref. [9] confirms the finding of a bound-state pole with $J = 0$ and $I = 0$ near and below the threshold of $\rho\rho$, which could be tentatively associated with the resonance $f_0(1370)$, as obtained previously in Ref. [68]. However, the deep pole found in the latter article for $J = 2$ and $I = 0$ is not confirmed in Ref. [9] because of the strong influence for these quantum numbers of the branch point singularity in $V(s)$ at $s = 3M_\rho^2$, which is neglected in Ref. [68].

The extension of the formalism of Ref. [9] for the uncoupled scattering to the coupled-channel case is given in Ref. [69]. The resulting formalism is applied to an $SU(3)$ study of vector–vector scattering. All the basic equations are completely analogous to the ones derived here [9], just by making use of a matrix language and including the appropriate subscripts to refer to the different PWAs coupled. The resulting T matrix of coupled PWAs is written as

$$T = D(s)^{-1}N(s) ,$$

(12.10)

with the matrices of functions

$$N_{ij}(s) = V_{ij}(s) ,$$

(12.11)

$$D_{ij}(s) = \gamma_{0;ij} + \gamma_{1;ij}(s - s_{th;j}) + \frac{1}{2}\gamma_{2;ij}(s - s_{th;j})^2$$

(12.12)

$$- \frac{(s - s_{th;j})s^2}{\pi}\int_{s_{th;j}}^{\infty}\frac{V_{ij}(s')\rho_j(s')ds'}{(s' - s_{th;j})(s')^2(s' - s)} .$$

The threshold for the jth channel is denoted by $s_{\text{th};j}$. The same strategy as before is followed and then the $D_{ij}(s)$ functions are matched with

$$\omega_{ij}(s) = \delta_{ij} + V_{ij}(s)g_j(s) + \frac{(s - s_{\text{th};j})s^2}{\pi} \int_{s_{\text{th};j}}^{\infty} \frac{V_{ij}(s')\rho_j(s')ds'}{(s' - s_{\text{th};j})(s')^2(s' - s)} . \quad (12.13)$$

at around $s_{\text{th};j}$. As noticed in Ref. [69] the LHC for $V_{ij}(s)$ is below the thresholds $s_{\text{th};i}$ and $s_{\text{th};j}$ and, thus, it makes sense to perform the Taylor expansion of $V_{ij}(s)$ (present in $\omega_{ij}(s)$) around $s_{\text{th};j}$. As a result we have the analogous expressions of Eq. (12.9), but now involving matrix indices. Thus,

$$\gamma_{0;ij} = 1 + V_{ij}(s_{\text{th};j})g_j(s_{\text{th};j}) , \quad (12.14)$$
$$\gamma_{1;ij} = \omega'_{ij}(s_{\text{th};j}) ,$$
$$\gamma_{2;ij} = \omega''_{ij}(s_{\text{th};j}) .$$

A set of poles with $J = 0$ and 1 of positive parity (only S-wave scattering is considered) results. For example, pole positions close to the $f_0(1370)$ and $f_0(1710)$ are reported. Similarly as in the $\rho\rho$-scattering study of Ref. [9], the coupled-channel one of Ref. [69] did not find any tensor resonance contrary to Refs. [68, 70], which only employ the contact interactions that result by freezing the vector propagators in the tree-level amplitudes obtained from the hidden-gauge chiral symmetry Lagrangian.

We would also like to stress that the matching procedure to determine the values of the subtraction constants that could appear in the N and D functions could also be made (if appropriate) by reproducing the perturbative T matrix at some particular energies. By assuming naive dimensional analysis one could determine as well the power counting of the subtraction constants by varying the position of the subtraction point and study how the subtraction constants run with the latter, as done in Ref. [64] for NN scattering and ChPT as EFT. In this way the mentioned matching with the perturbative T matrix could also be done algebraically.

The first-iterated solution of the N/D method has the virtue that its does not generate spurious LHC contributions in the $D(s)$ function, which would affect $T(s)$ as well. This might happen when coupling channels with different masses by using the perturbative approximation for the matrix $\mathcal{N}$, Eq. 10.2, as well as in the IAM [35, 71, 72]. In this way, by using the first-iterated solution of the N/D method, one can avoid the presence of such artifacts when looking for the pole positions of resonances (in the unphysical RSs) and bound states (in the physical RS). These poles correspond to zeros of $\det D$ in the appropriate RS. Of course, this potential problem is also cured by fully solving the N/D method in terms of the given $\Delta(p^2)$, as in Chap. 11, for the case of coupled channels. This is done in Refs. [7, 64–66] for studying NN scattering.

Chapter 13
Final(Initial)-State Interactions.
Unitarity and Analyticity Requirements.
Watson Final-State Theorem

In order to test the dynamics of a system it is common to use probes that only interact feebly with the system, which dynamics is driven by stronger interactions. The former weaker process can then drive to different final states that interact strongly among them. One then typically distinguish the triggering mechanism of the weak reaction from the afterwards strong rescattering processes, due to the strong interactions between possible different final states. The later are called final-state interactions (FSI) By time-reversal invariance, we could also have strong interaction processes driving to feeble probes in the final state, in this case we would call them as initial-state interactions. Unless the opposite is stated we refer in the subsequent to FSI, though it should be kept in mind that the formalism developed is also applicable to the initial-state interacting case. Examples are $e^+ e^- \leftrightarrow \pi\pi$, $\gamma N \leftrightarrow \pi N$, $J/\Psi \rightarrow \gamma\pi\pi$, etc.

The total T and S matrices comprise the weaker and strong interacting processes. These matrices are related in the usual manner by Eq. (1.6). For example, the S matrix for $\gamma\gamma \rightarrow \pi\pi$, $K\bar{K}$ (we consider that the states $|\pi\pi\rangle$ and $|K\bar{K}\rangle$ have some definite isospin I, which is conserved the strong interactions), may be represented by a 3×3 S matrix (without further specification of helicity and momentum arguments because we have in mind some definite partial-wave projection),

$$S = \begin{pmatrix} S_{\gamma\gamma \to \pi\pi} & S_{\pi\pi \to \pi\pi} & S_{K\bar{K} \to \pi\pi} \\ S_{\gamma\gamma \to K\bar{K}} & S_{\pi\pi \to K\bar{K}} & S_{K\bar{K} \to K\bar{K}} \\ S_{\gamma\gamma \to \gamma\gamma} & S_{\pi\pi \to \gamma\gamma} & S_{K\bar{K} \to \gamma\gamma} \end{pmatrix} . \tag{13.1}$$

The unitarity of the S matrix implies the relations of Eqs. (1.7) and (1.8) Keeping only terms linear in the weaker interactions (whose matrix element for producing the channel i is indicated by F_i and generically designated as form factor), we have from unitarity that

$$F_i - F_i^\dagger = i \sum_j \int dQ_j \theta_j F_j T_{ji}^\dagger , \tag{13.2}$$

© The Author(s), under exclusive licence to Springer Nature Switzerland AG 2019
J. A. Oller, *A Brief Introduction to Dispersion Relations*, SpringerBriefs in Physics,
https://doi.org/10.1007/978-3-030-13582-9_13

where only states that are open contribute to the sum, $\theta_j = \theta(s - s_{th;j})$. This relation is valid because of extended unitarity above the lightest threshold open [2], even if some of the final states $|i\rangle$ are closed.

As usual, the unitarity relation becomes simpler when decomposing the T matrix elements in PWAs. For instance, for the pion form factor corresponding to the transition $\langle\gamma|T|\pi^+\pi^-\rangle$ only the $J = 1$ $\pi\pi$ PWA contributes,

$$\langle\gamma(q)|T|\pi^+(p)\pi^-(p')\rangle = e\varepsilon(q)_\mu(p - p')^\mu F_{\pi\pi}(s) \tag{13.3}$$

We also concentrate ourselves with two-body interactions in the final state and assume time-reversal symmetry, so that the PWAs are symmetric. As a result the unitarity relation of Eqs. (1.7) and (1.8) between state with definite total spin J, ℓ and S simplifies as,[1]

$$\Im F_i(s) = \sum_j F_j(s)\rho_j(s)\theta_j T_{ij}(s)^* \tag{13.4}$$

$$= \sum_j F_j(s)^*\rho_j(s)\theta_j T_{ij}(s) ,$$

where the subscripts i, j denote the different partial-wave projected states.

For the uncoupled case the unitarity relation of Eq. (13.4) already implies an important result known as the Watson final-state theorem. It states that the phase of $F(s)$ along the RHC and up to the next higher threshold is the same as the phase of $T(s)$ modulo π. This follows because from Eq. (13.4) we have in the one-channel case that

$$\Im F_1(s) = F_1(s)T_{11}(s)^*\rho_1(s)\theta_1 . \tag{13.5}$$

Since the lhs is real so must be the rhs and from this observation it follows the Watson final-state theorem.

Its generalization to coupled channel proceeds in the following manner. Let us employ Eq. (7.2) to express the T matrix of PWAs in terms of the matrices $\mathcal{N}$ and $g(s)$. We also rewrite the lhs of Eq. (13.4) as $(F_i(s) - F_i(s)^*)/2i$ and group together the $F_i(s)^*$ on the rhs. It results

$$F_i(s) = \sum_j \left[\delta_{ij} + 2i\rho_j(s)\theta_j T_{ij}(s)\right] F_j(s)^* . \tag{13.6}$$

It is convenient to write this expression in matrix notation and extract to the left $T(s) = (\mathcal{N}^{-1} + g)^{-1}$,

---

[1]Depending on the nature of the initial state, i.e., the probe, it might be necessary to decompose it as well in PWAs. For example, for $\gamma\gamma$ to meson–meson the reader could consult Refs. [73, 74] where the $\gamma\gamma$ state is decomposed in helicity amplitudes.

$$F = \left(\mathcal{N}^{-1} + g\right)^{-1} \left(\mathcal{N}^{-1} + g + 2i\rho(s)\theta\right) F^* . \tag{13.7}$$

Notice that the matrix $\rho(s)$ is introduced already in Eq. (6.27) and it includes the matrix of $\theta(s)$ Heaviside functions $\theta_i(s)$. This is consistent with the expression in Eq. (13.7) since $\rho\theta = \rho$. Taking into account that $g(s) + 2i\rho(s)\theta(s) = g(s)^*$, one obtains the following result valid along the RHC,

$$\left(\mathcal{N}^{-1} + g\right) F = \left(\mathcal{N}^{-1} + g^*\right) F^* . \tag{13.8}$$

Multiplying by $\mathcal{N}$ on both sides, we obtain the relation

$$[I + \mathcal{N}(s)g(s)] F(s) = \left[I + \mathcal{N}(s)g(s)^*\right] F(s)^* . \tag{13.9}$$

This is the generalization of the Watson final-state theorem to coupled channels. Notice that in the uncoupled case the phase of $1 + \mathcal{N}g$ is minus the phase of the PWA and, since Eq. (13.9) requires $(1 + \mathcal{N}g)F$ be real, it follows the Watson final-state theorem.

An important consequence from Eq. (13.9) is that

$$[I + \mathcal{N}(s)g(s)] F(s) \tag{13.10}$$

is free of RHC [75], since it is the same as its complex conjugate for s above the lightest threshold. Notice also that $\mathcal{N}$ has no RHC and we could have taken its complex conjugate in Eq. (13.9). As a result we can write $F(s)$ as

$$F(s) = [I + \mathcal{N}(s)g(s)]^{-1} L(s) , \tag{13.11}$$

where $L(s)$ is an $n \times 1$ column vector, with n the number of coupled PWAs. The characteristic feature of $L(s)$ is that it has no RHC, as follows from the previous discussion. Therefore, it only has LHC.

These results also apply if instead we used the N/D method to write $T(s)$ as $D^{-1}N$ in Eq. (13.6). By following similar steps, we would have instead of Eq. (13.7) the relation

$$F(s) = D^{-1}(s)[D + N2i\rho\theta] F(s)^* = D^{-1}(s)D(s)^* F(s)^* , \tag{13.12}$$

so that,

$$D(s)F(s) = D(s)^* F(s)^* , \tag{13.13}$$

where we have taken into account that $\Im D(s) = -N(s)\rho$. Thus, we can also conclude that $F(s)$ can be written as

$$F(s) = D(s)^{-1}L(s) , \qquad (13.14)$$

with $L(s)$ having at most LHC (if any). This expression is even more convenient than that of Eq. (13.11) because the function $D(s)$ has only RHC, while $I + \mathcal{N}(s)g(s)$ has RHC and LHC (as $\mathcal{N}(s)$ has LHC and $g(s)$ has RHC).

As an example, if the vector form factor, Eq. (13.3), is expressed as in Eq. (13.14), the function $L(s)$ is free of any cut because the form factor $F(s)$ has only RHC. The latter statement follows because the Mandelstam variable s is the only Lorentz invariant that can be made out of the momenta of the two on-shell pions (as for each pion $p_i^2 = m_\pi^2$ and $p_1 p_2 = s/2 - m_\pi^2$).

For the two-coupled channel case from the unitarity relations of Eq. (13.4) one can actually express $F_2(s)$ in terms of $F_1(s)$ above the higher threshold [75, 76]. This is not surprising since unitarity implies two equations and $F_2(s)$ comprises two real functions (its real and imaginary parts). Following this logic, we could then expect quite generally that in a problem with $2n$ or $2n + 1$ coupled PWAs we could express n of the form factors in terms of the other n or $n + 1$ ones. Being more explicit for the two-coupled channel case we have for the strong S matrix the following parameterization that guarantees unitarity,

$$S(s) = \begin{pmatrix} \eta e^{2i\delta_1} & i\sqrt{1-\eta^2}e^{i(\delta_1+\delta_2)} \\ i\sqrt{1-\eta^2}e^{i(\delta_1+\delta_2)} & \eta e^{2i\delta_2} \end{pmatrix} . \qquad (13.15)$$

We take into account the relation between the S and T matrices given in Eq. (2.52), to express the latter in terms of $\eta \equiv \cos 2\alpha$ ($\sin 2\alpha = \sqrt{1-\eta^2}$) and the phase shifts δ_i. Next, we write the form factors as

$$F_i(s) = f_i(s)e^{i(\delta_i(s)+\phi_i(s))} , \qquad (13.16)$$

and take the real and imaginary parts of the unitarity relations for the form factors written as in Eq. (13.6). One then has the following relations equivalent to the two unitarity constraints,

$$(1 - \cos 2\alpha) \cos \phi_1 \, f_1 = \sqrt{\frac{\rho_2}{\rho_1}} \sin 2\alpha \sin \phi_2 f_2 , \qquad (13.17)$$

$$(1 + \cos 2\alpha) \sin \phi_1 \, f_1 = \sqrt{\frac{\rho_2}{\rho_1}} \sin 2\alpha \cos \phi_2 f_2 .$$

Dividing the former by the latter equation implies that

$$\tan \phi_1 \tan \phi_2 = \tan^2 \alpha . \qquad (13.18)$$

Adding the squared of them, we obtain

$$\frac{\rho_2 f_2^2}{\rho_1 f_1^2} = \tan^2 \alpha + \frac{4 \cos 2\alpha}{\sin^2 2\alpha} \sin^2 \phi_1 \, . \tag{13.19}$$

Making use of Eqs. (13.18) and (13.19) one can then obtain f_2 and ϕ_2 in terms of f_1 and ϕ_1 (or viceversa).

Chapter 14
The Omnès Solution. Reasoned Warnings on the Use of the Omnès Function

We consider along this section that the uncoupled unitarity relation of Eq. (13.5) can be applied, at least in good approximation, and assume that the strong interacting PWA is known. Given a form factor with only RHC, like for example the vector or scalar form factors of two hadrons, the unitarity relation, Eq. (13.5), provides us with its discontinuity along this cut. The phase of the form factor $F(s)$, $\varphi(s)$, is the same as the phase of the PWA $T(s)$ [also denoted then by $\varphi(s)$] because of the Watson final-state theorem (we are here suppressing any subscript).[1] In the strict elastic region $\varphi(s) = \delta(s)$ but, as we show below, it might be that still the phase of the form factor corresponds approximately to that of the PWA, while the latter departs strongly from $\delta(s)$ in a region with marked inelasticity. The reason is that the form factor mostly couples to a given eigen-channel that diagonalizes the S matrix, for which the elastic treatment holds.

The solution for an analytical function in the cut complex s plane, with a branch point singularity at s_{th} associated with a RHC, along which its phase is known, can be written in terms of the so-called Omnès function. The idea is relatively straightforward and can be implemented in two steps.

First, by the knowledge of $\varphi(s)$ we construct an analytical function with a RHC and branch point discontinuity at s_{th} by writing down the DR

$$\omega(s) = \sum_{i=0}^{n-1} a_i s^i + \frac{s^n}{\pi} \int_{s_{th}}^{\infty} \frac{\varphi(s')ds'}{(s')^n(s'-s)} ,\qquad(14.1)$$

where we have assumed that $\varphi(s)$ does not diverge stronger than s^{n-1} for $s \to \infty$, with n a finite integer. We have introduced n subtraction constants so that the result is independent of the subtraction point. Along the RHC this function fulfills that $\omega(s + i\varepsilon) - \omega(s - i\varepsilon) = 2i\varphi(s)$. Second, we next define the Omnès function, $\Omega(s)$, as

---

[1] If there is a difference between these two phases of π then just take $-F(s)$.

© The Author(s), under exclusive licence to Springer Nature Switzerland AG 2019
J. A. Oller, *A Brief Introduction to Dispersion Relations*, SpringerBriefs in Physics,
https://doi.org/10.1007/978-3-030-13582-9_14

$$\Omega(s) = \exp \omega(s) . \tag{14.2}$$

We always have the freedom to normalize the Omnès function such that $\Omega(0) = 1$, which fixes $a_0 = 1$. It follows also that the combination

$$R(s) = \frac{F(s)}{\Omega(s)} , \tag{14.3}$$

is real for $s > s_{\text{th}}$ and it has no cuts, so that it is a meromorphic function of s in the first RS of the whole complex s plane.

Let us consider first that $\omega(s)$ is finite along the RHC, so that $0 < |\Omega(s)| < \infty,$[2] and there are no bound states (i.e., $F(s)$ has no poles). It is known in complex analysis that any function that is analytic in the whole complex s plane is constant or unbounded. If we apply this theorem to $R(s)$, we learn then that $F(s)$ diverges as much as or stronger than $\Omega(s)$ for $s \to \infty$. As result we conclude in this case that we can express $F(s)$ as

$$F(s) = R(s)\Omega(s) , \tag{14.4}$$

with $R(s)$ a constant or an analytical function which is unbounded at infinity. Indeed, we can expect exponential divergences in $\Omega(s)$ from Eq. (14.2) when the DR for $\omega(s)$ requires for convergence more than one subtraction. The conclusion follows by a similar analysis as the one performed between Eqs. (4.10) and (4.12) in relation with the Sugawara–Kanazawa theorem. Thus, if $\varphi(s)/s^{n-1}$ ($n \geq 2$) were not vanishing for $s \to \infty$, one would have logarithmic divergences like $s^{n-1} \log s$ (here there is only RHC). These divergences could not be canceled by the $a_n s^{n-1}$ term. Therefore, $R(s)$ would be an exponential function so as to guarantee that $F(s)$ does not diverge stronger than a power of s for $s \to \infty$ (and it is then amenable for a DR). Regarding this point, one would expect that a hadronic form factor would typically vanish for $s \to \infty$ because of the finiteness of the non-perturbative scale of QCD, Λ_{QCD}, as also suggested by the quark counting rules [77–79], and then being amenable to a DR. By the same token, one would also expect intuitively that the phase of the form factor tends to a constant limit for $s \to \infty$. However, these expectations could fail in the case of singular interactions at the origin.

We can say more about $R(s)$. Let $P(s)$ be the polynomial made out of the possible zeroes of $F(s)$ (if any), and let $Q(s)$ be another polynomial whose zeroes are the possible poles (if there exists any bound state) of $F(s)$. Next, we multiply $F(s)$ by the rational function $Q(s)/P(s)$ and perform a DR of the analytical function $\log [F(s)Q(s)/P(s)]$ in the cut complex s plane circumventing the RHC. The discontinuity of this function along the RHC is $2i[\delta(s + i\varepsilon) - \delta(s - i\varepsilon)]$, the one corresponding to function $\omega(s)$. Of course, we are assuming also here that $F(s)$ has a finite number of zeroes and bound states (these numbers are p and q, respectively) and that $\log F(s)Q(s)/P(s)$ is amenable to a DR treatment. There, this procedure implies

---

[2]Later we discuss a specific situation when this is not the case.

that $F(s)$ can be written as in Eq. (14.4) with $R(s) = Q(s)/P(s)$. In the subsequent we further require that $F(s)$ resulting from this analysis does not grow exponentially, so that we are driven to admit that a once-subtracted DR is possible for $\omega(s)$ in Eq. (14.1). In other terms, we assume that for $s \to \infty$ the ratio $|\varphi(s)/s| < s^{-\gamma}$ for some $\gamma > 0$, because otherwise we could apply the analysis above below Eq. (14.4), and based on the process followed for the demonstration of the Sugawara–Kanazawa theorem.[3] We then arrive to the following expression for $F(s)$ that we consider in the following:

$$F(s) = \frac{P(s)}{Q(s)} \Omega(s) , \tag{14.5}$$

$$\Omega(s) = \exp \omega(s) , \tag{14.6}$$

$$\omega(s) = \frac{s}{\pi} \int_{S_{th}}^{\infty} \frac{\varphi(s')ds'}{s'(s' - s)} . \tag{14.7}$$

Here, $P(s)$ absorbs the required normalization constant to permit our choice $\Omega(0) = 1$ without loss of generality.

Let us work out the behavior of $\Omega(s)$ in the limit $s \to \infty$ by taking for granted the existence of the limit $\varphi(\infty) < \infty$. For that we again proceed as in Eq. (4.10) and then we decompose $\omega(s)$ in Eq. (14.7) as

$$\omega(s) = \varphi(\infty) \frac{s}{\pi} \int_{S_{th}}^{\infty} \frac{ds'}{s'(s' - s)} + \frac{s}{\pi} \int_{S_{th}}^{\infty} \frac{\varphi(s') - \varphi(\infty)}{s'(s' - s)} ds' . \tag{14.8}$$

Thus, for $s \to \infty$ we have

$$\omega(s + i\varepsilon) \xrightarrow[s \to \infty]{} -\frac{\varphi(\infty)}{\pi} \log \frac{s}{S_{th}} + i\varphi(\infty) - \frac{1}{\pi} \int_{S_{th}}^{\infty} \frac{\varphi(s') - \varphi(\infty)}{s'} ds' , \tag{14.9}$$

and the logarithmic divergence is the one that dominates for $s \to \infty$. The other two terms in the previous equation are constant ones, the first one is purely imaginary and gives the phase of $\Omega(s)$ while the latter is a constant stemming from the second integral in Eq. (14.8) that renormalizes $P(s)$ in the considered limit of $s \to \infty$. As a result, we have for $\Omega(s)$ the limit behavior

---

[3] In nonrelativistic scattering we know from the Levinson theorem [17, 80] that $\delta(0) - \delta(\infty) = (n + q/2)\pi$, where n is the number of bound states in the problem and q only applies to S-wave ($\ell = 0$), being the number of zero energy S-wave resonances. For the precise condition of this later case consider Eq. (95) of Ref. [17].

$$\Omega(s) \xrightarrow[s\to\infty]{} C_\Omega \, e^{i\varphi(\infty)} \times \left(\frac{s_{th}}{s}\right)^{\frac{\varphi(\infty)}{\pi}} . \tag{14.10}$$

This translates into the form factor $F(s)$, Eq. (14.5), as

$$F(s) \xrightarrow[s\to\infty]{} C_F \, e^{i\varphi(\infty)} \times s^{p-q-\frac{\varphi(\infty)}{\pi}} , \tag{14.11}$$

where C_Ω and C_F are constants.

The Eq. (14.11) offers interesting corollaries
(i) If the high-energy behavior of $F(s)$ is considered to be known and it is of the form s^ν, then we have from this equation that

$$p - q - \frac{\varphi(\infty)}{\pi} = \nu , \tag{14.12}$$

which is a kind of relativistic Levinson theorem for the form factor.
(ii) When modeling interactions with limited information, so that we are able to achieve some partial control on the PWA and form factor, we should keep constant under variation of the parameters the relation of Eq. (14.12). Since ν is fixed then we would require that

$$p - q - \frac{\varphi(\infty)}{\pi} = \text{fixed} \tag{14.13}$$

as the parameters vary. In this way, if, e.g., $\varphi(\infty)/\pi$ decreases by one unit and there are no bound states in the system then we should introduce an extra zero in the form factor, so that p increases by one compensating unit. A similar logic would apply for other possible situations.

(iii) We should stress that while we can compensate for the strong-model effects discussed in (ii), by increasing/decreasing p, q and $\varphi(\infty)/\pi$, this is not possible for $\Omega(s)$, which then could be driven into a very troublesome behavior. That is, $\Omega(s)$ is expected to be more strongly dependent on fine details of the hadronic model and it should be used with care, e.g., within a formula like that for $F(s)$ in Eq. (14.5).

As an important example that illustrates the previous points (i)–(iii), we refer to the pion scalar form factor, associated with the light-quark scalar source $\bar{u}u + \bar{d}d$, which is defined as

$$F(s) = \int d^4x \, e^{i(p+p')x} \langle 0|m_u\bar{u}(x)u(x) + m_d\bar{d}(x)d(x)|0\rangle . \tag{14.14}$$

Here u and d are the up and down quarks, m_u and m_d are the masses of these quarks, in order, and $s = (p + p')^2$. In the following we consider the isospin limit (nominally, $m_u = m_d$).

The FSI for this form factor are driven by the isoscalar scalar $\pi\pi$ interaction, which was discussed for low energies with its salient feature of the appearance of the

$f_0(500)$ or σ meson in Chap. 8. Another phenomenologically relevant channel is the $K\bar{K}$ one, with a threshold at 991.4 MeV [34]. Apart from the $f_0(500)$ or σ resonance there is also the $f_0(980)$ resonance, which is relatively narrow [34] and it manifests as a steep rise of the isoscalar scalar phase shifts at around the two-kaon threshold. This resonance couples much more strongly to $K\bar{K}$ than to $\pi\pi$ [81], which causes that as soon as the $K\bar{K}$ is open there is an active conversion of pionic flux in a kaonic one. As a result, the inelasticity parameter η_1 rapidly drops from 1 below the $K\bar{K}$ threshold to much smaller values for $\sqrt{s} > 2m_K$. The aforementioned rapid rise of the phase of the isoscalar scalar $\pi\pi$ PWA $T(s)$, $\varphi(s)$, could be abruptly interrupted at the $K\bar{K}$ before it reached π degrees. All depends on whether $\delta(s)$ at $s = s_K = (2m_K)^2$ is larger or smaller than π, which might be easily changed within the parameters of the hadronic model, being both situations compatible with the present experimental phase shifts at around $s = 1$ GeV^2. As a result, the Omnès function for this case would have two dramatically different behaviors under tiny changes of the parameters, depending on whether $\delta(s_K)$ is larger or smaller than π. In the former case $\Omega(s)$ is huge at the point where $\delta(s) = \pi$ (this point is below s_K), while for the later case $\Omega(s)$ is nearly zero just below the $K\bar{K}$ threshold. This pathological situation was discussed in great detail in Ref. [82]. We also refer to Refs. [10, 81] for explicit accounts of the mentioned experimental data for the isoscalar scalar meson–meson interactions.

Let us exemplify this situation by performing an explicit calculation by identifying $\varphi(s)$ with the phase of the PWA $T(s)$ along the RHC. For the numerical evaluation of the DR for $\omega(s)$, Eq. (14.7), it is convenient to rewrite it so as to avoid the explicit numerical calculation of the Cauchy principal value of the integral involving $\varphi(s')$. We then have

$$\omega(s) = \varphi(s)\frac{s}{\pi}\int_{s_{th}}^{\infty}\frac{ds'}{s'(s'-s)} + \frac{s}{\pi}\int_{s_{th}}^{\infty}\frac{\varphi(s')-\varphi(s)}{s'(s'-s)}ds' , \qquad (14.15)$$

and the former integral can be evaluated algebraically.

The situation in which $\delta(s_K) \to \pi$ drives to a singularity in the Omnès function $\Omega(s)$. When this happens, with a subtraction constant around -2.45, the phase of the strong PWA becomes discontinuous for s above s_K. We plot $\delta(s)$ in the left top panel of Fig. 14.1 and $\varphi(s)$ in the right top one. The PWA $T(s)$ in terms of the phase shifts $\delta(s)$ is given by

$$T(s) = |T(s)|e^{i\varphi(s)} = \frac{1}{2\rho}[\eta\sin 2\delta + i(1 - \eta\cos 2\delta(s))] . \qquad (14.16)$$

Thus, when $\delta(s_K) < \pi$ we have that above s_K the real part of $T(s)$ changes sign (since $\delta(s)$ keeps growing), and then the phase of $T(s)$ experiences a rapid decrease from values near π below s_K to values in the interval $[0, \pi/2]$ (the imaginary part of $T(s)$ is always positive because of unitarity, $\eta \leq 1$). This transition in $\varphi(s)$ from values near to π to others below $\pi/2$ becomes more abrupt as $\delta(s_K) \to \pi^-$, and in reaching this limit the phase $\varphi(s)$ becomes discontinuous at s_K. On the other hand, when $\delta(s_K) > \pi$ the function $\varphi(s)$ keeps growing because the real part of $T(s)$ does not change sign and when η becomes small then it is clear that the imaginary part of $T(s)$ becomes larger than the real part $[\varphi(s) > \pi$ for $s > s_K]$. The presence of such a discontinuity in $\varphi(s)$ at s_K by an amount of $\pi/2$ drives to a singularity in $\omega(s)$ at $s = s_K$. This singularity is of the end-point type, as it is clear by splitting the integral for $\omega(s)$ in two parts, from s_π ($s_\pi = 4m_\pi^2$) to s_K and from the latter to ∞, with $\varphi(s_K - \varepsilon) - \varphi(s_K + \varepsilon) = \pm\pi/2$. In the latter expression, the plus sign applies when $\delta(s_K - \varepsilon) \to \pi^-$ and the minus sign when $\delta(s_K - \varepsilon) \to \pi^+$. The resulting logarithmic singularity in $\omega(s)$ stems then from the fact that the Cauchy's principal value of the integral around $s = s_K$ does not get rid of the pole singularity in the integrand from the factor $1/(s' - s)$. Thus, we are driving to the divergence

$$\frac{1}{\pi}\left[\int^{s_K - \Delta} \frac{\varphi(s_K - \varepsilon)ds'}{s' - s_K} + \int_{s_K + \Delta} \frac{\varphi(s_K + \varepsilon)ds'}{s' - s_K}\right]$$
$$\to \frac{1}{\pi}[\varphi(s_K - \varepsilon) - \varphi(s_K + \varepsilon)]\log\Delta = \pm\frac{1}{2}\log\Delta, \qquad (14.17)$$

with $\Delta \to 0^+$ and $\delta(s_K - \varepsilon) \to \pi^\mp$, in order. In this way, when exponentiating $\omega(s)$ to get $\Omega(s)$ this divergent contribution in the exponent gives rise to $(\sqrt{\Delta})^{\pm 1}$. Therefore, $\Omega(s)$ has finally a pole when $\delta(s_K) \to \pi^+$ and a zero when $\delta(s_K) \to \pi^-$. This behavior is represented in the right bottom panel in Fig. 14.1.

This pathological situation has a reflection in the condition expressed in Eq. (14.13), because there is a jump by one in $\varphi(\infty)/\pi$ between the two situations $\delta(s_K - \varepsilon) \to \pi^\pm$. Thus, imposing continuity in the transition $\delta(s_K - \varepsilon) < \pi$ to $> \pi$ requires that p increases by 1, that is, there should be one more zero for $\delta(s_K) > \pi$ as compared with the opposite situation. If we would require the continuity from $\delta(s_K) > \pi$ to $< \pi$ we would have to increase q by one and had bound sate (a pole in the first Riemann sheet). This latter situation can be ruled out in pion physics. It follows then that an Omnès representation of the isoscalar scalar $\pi\pi$ PWA in the case $\delta(s_K) > \pi$ requires the function $\Omega(s)$ to have a zero at the point at which $\Im T(s) = 0$ for $s < s_K$. Well, applying an Omnès representation for $T(s)$ itself this is also a consequence of unitarity because $T(s) = e^{i\delta}\sin(\delta)/\rho$ in the elastic region below the $K\bar{K}$ threshold.

Similar reasoning was applied in Ref. [82] to the pion scalar form factor $F(s)$ which follows (in good approximation) the phase of the isoscalar scalar $\pi\pi$ PWA $T(s)$ (even somewhat above the $K\bar{K}$ threshold). This is shown by explicit calculations of $F(s)$ within other approaches [11, 83, 84]. Indeed, such a situation might be expected by realizing that the $f_0(980)$ couples much more strongly to kaons than to pions, e.g., Ref. [81] reports that the coupling to kaons is larger by a factor 3. As a result, the admixture between the pion and kaon channels is suppressed and both of them follow their own eigen-channel of the isoscalar scalar meson–meson PWAs. We refer to Refs. [82, 85] for detailed discussions that provide the explicit expression for the eigen-channels and eigen-phases.

We would also like to mention that one can precisely determine the point $s = s_1$ at which the form factor has a zero when $\delta(s_K) > \pi$ as find out in Ref. [82]. This reference writes down a twice-subtracted DR for the form factor,

$$F(s) = F(0) + \frac{1}{6}\langle r^2\rangle_s^\pi s + \frac{s^2}{\pi} \int_{s_\pi}^\infty \frac{\Im F(s')ds'}{(s')^2(s'-s)} . \tag{14.18}$$

In this expression, $\langle r^2\rangle_s^\pi$ is the quadratic scalar radius of the pion. Indeed one expects from asymptotic QCD [86] that $F(s)$ vanishes at infinity so that the written DR should converge fast, which is of particular interest for relatively low energies. It is then clear from the integral representation of $F(s)$ in Eq. (14.18) that the only point at which $F(s)$ can vanish for $s < s_k$ is where $\Im F(s) = 0$ (since the subtraction polynomial in Eq. (14.18) is real). The latter fact can only occur when $\delta(s) = \pi$ since there is only one zero at such energies and $|\Im F(s)| = |F(s)\sin\delta(s)|/\rho(s)$ in the elastic region, $s < s_K$, and $\delta(s_K) > \pi$. This in turns fixes the first order polynomial that should multiply the Omnès function $\Omega(s)$ so as to achieve a continuous transition for $\delta(s_K)$ greater or smaller than π.

In summary, one should better use the function

$$\Omega(s) = \begin{cases} \exp\omega(s) & , \ \delta(s_K) < \pi , \\ \frac{s_1-s}{s_1}\exp\omega(s) & , \ \delta(s_K) > \pi . \end{cases} \tag{14.19}$$

A clear lesson that follows from the discussion in this section is that one should use an Omnès function with great care when employing it while doing fits to data. The latter requires varying the parameters of the theory and one should avoid possible instable behaviors associated with rapid movements in the phases integrated that could strongly affect an Omnès function. As we have seen, nonsense results could arise by a nearby discontinuity in the space of parameters. The fulfillment of the requirement in Eq. (14.13) should then be pursuit, and for the phase of the isoscalar scalar $\pi\pi$ PWA one should use the function in Eq. (14.19) instead of a pure Omnès function, cf. Eq. (14.6).

Given a form factor which also involves LHC,[4] e.g., that for $\gamma\gamma \to \pi\pi$, we could also define the function $R(s)$ as in Eq. (14.3), although now this function also contains LHC, and then we denote it by $L(s)$ [analogously to Eq. (13.14)]. Nonetheless, the introduction of this function allows a clear splitting between the RHC and LHC contributions that is also exploited in the literature. One typically writes down a DR for $L(s)$ along the LHC,

$$L(s) = \sum_{i=1}^{n-1} a_i s^i + \frac{s^n}{\pi} \int_{-\infty}^{s_L} \frac{\Im L(s')ds'}{(s')^n(s'-s)} , \tag{14.20}$$

$$F(s) = \Omega(s)L(s) ,$$

$$\Im L(s) = \Omega(s)^{-1}\Im F(s) , \ s < s_L ,$$

---

[4]Maybe some readers are used to consider that the form factors should only have RHC. Here we use the notation introduced in Chap. 13.

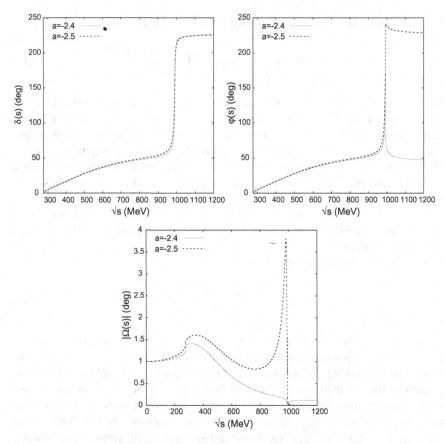

Fig. 14.1 From left to right and top to bottom: Phase shifts, $\delta(s)$, phase of $T(s)$, $\varphi(s)$, and the Omnès function, $|\Omega(s)|$. The solid line corresponds to the subtraction constant a of the $g(s)$ function with the value $a = -2.4$ and the dashed line to $a = -2.5$. We have modeled the $\pi\pi$ and $K\bar{K}$ channels with $I = \ell = 0$ by unitarizing the lowest order ChPT amplitudes. We use Eq. (8.29) and the 2×2 matrix $\mathcal{N}(s)$ is identified with the leading ChPT amplitudes, Eq. (8.30)

with s_L the upper limit for the LHC. In this way, the Omnès function can be known (at least partially) from the knowledge of the strong PWAs along the RHC and then one needs to know $\Im F(s)$ along the LHC. Of course, in the pure elastic case the phase of the Omnès function is the phase of the strong PWA $T(s)$ and we could proceed as discussed above in this section. For the particular case of $\gamma\gamma \to \pi^0\pi^0$ its S wave contribution is discussed in Refs. [87, 88]. The subtraction constants can be adjusted by employing the Low's theorem, which implies that for $s \to 0$ the total $F(s)$ tends to its renormalized Born term contribution (involving the values of the physical couplings and masses). The other subtraction constant is fixed by matching with the one-loop ChPT calculation of Refs. [89, 90]. One could approach $\Im F(s)$ by the contributions from the Born terms and the crossed exchanges of the J^{PC}

resonance multiplets 1^{--} and 1^{++} as in Ref. [88], where explicit formulas for the resonance-exchange tree-level amplitudes can be found.

Incidentally, Refs. [87, 88] use a somewhat different unitarization procedure than Eq. (14.20) to calculate the low-energy cross section for $\gamma\gamma \to \pi^0\pi^0$. These references consider only S wave ($\pi^0\pi^0$ does not have P wave because of Bose–Einstein symmetry) and two isospin channels are possible, the isoscalar and the isotensor ones. The first step is to build up a function with only RHC by subtracting to $F_I(s)$ a function $\tilde{L}_I(s)$ that contains its LHC. Namely, the new function is

$$\mathcal{F}_I(s) = \frac{F_I(s) - \tilde{L}_I(s)}{\Omega_I(s)} . \tag{14.21}$$

Next, Refs. [87, 88] perform a twice-subtracted DR for the latter, in terms of which $F_I(s)$ reads

$$F_I(s) = \tilde{L}_I(s) + a_I \Omega_I(s) + c_I s \Omega_I(s) + \Omega_I(s) \frac{s^2}{\pi} \int_{4m_\pi^2}^{\infty} \frac{\tilde{L}_I(s') \sin\varphi_I(s') ds'}{(s')^2 (s' - s)|\Omega_I(s')|} . \tag{14.22}$$

The subtraction constants are fixed as explained above by considering the Low's theorem and matching with the one-loop ChPT calculations of $\gamma\gamma \to \pi\pi$ in Refs. [89, 90]. At the practical level $\tilde{L}_I(s)$ is also approximated in Refs. [87, 88] by the tree-level amplitudes including the Born terms and the exchange of the 1^{--} and 1^{++} multiplets of vector and axial resonances, in order. In this way, the Low's theorem requires that

$$\lim_{s \to 0} \left[F_I(s) - \tilde{L}_I(s) \right] = \mathcal{O}(s) , \tag{14.23}$$

from which it follows that $a_I = 0$ in Eq. (14.22). Indeed, the contributions of the 1^{++} axial resonances are more important at low energies than that of the 1^{--}. Actually, the former appear one order lower in the chiral expansion than the latter. Despite that the explicit axial exchanges are neglected in Ref. [87], while they are taken into account in Ref. [88]. The calculations performed in this reference confirm that these contributions are phenomenologically relevant and should not be neglected since their contributions are around a 30% of the full result. A major step forward of Ref. [88] compared to Ref. [87] is to use the stable $\Omega_0(s)$ function as defined in Eq. (14.19), instead of just a pure Omnès function. In this way, the output at low energies is much more stable under changes of the parameterizations used for the isoscalar scalar $\pi\pi$ phase shifts in the region of the $f_0(980)$ resonance, accomplishing a reduction of about a factor of 2 in the uncertainty of the cross section for $\gamma\gamma \to \pi^0\pi^0$ at around the mass of the $\rho(770)$, and about a 25% already at around $\sqrt{s} = 500$ MeV. Notice, that even if for $\delta_0(s_K) > \pi$ one has a zero in the denominator because $\Omega_0(s_1) = 0$, as defined in Eq. (14.19), the ratio $\sin \varphi_0(s')/|\Omega_0(s')|$ in the integrand of Eq. (14.22) is well defined because the zero of $\Omega_0(s')$ happens at the same point s_1 at which $\varphi_0(s_1) = \pi$, cf. Eq. (14.18).

Chapter 15
The Muskhelishvili-Omnès Problem in Coupled-Channel form Factors

The basic problem that we consider in this section is to find the possible solutions for a set of form factors $F_i(s)$, $i = 1 \ldots n$, ordered in increasing value of their thresholds $s_{\text{th};i}$. Each of the $F_i(s)$ has LHC for $s < s_L$ and RHC for $s > s_{\text{th}}$, where s_{th} is the lightest of all the thresholds $s_{\text{th};i}$ involved and s_L was defined above.

Along the RHC the imaginary part of $F_i(s)$ is given by the unitarity relation expressed in Eq. (13.4). The latter also allows us to know the discontinuities of these form factors along the RHC because they fulfill the Schwarz reflection principle,

$$F_i(s^*) = F_i(s)^* , \qquad (15.1)$$

since the form factors are real in the interval $s_L < s < s_{\text{th}}$. As a result, the discontinuity of $F_i(s)$ along the RHC obeys

$$\Im F_i(s + i\varepsilon) - \Im F_i(s - i\varepsilon) = 2i\Im F_i(s + i\varepsilon) , \ s > s_{\text{th}} . \qquad (15.2)$$

On the other hand, the discontinuity of these functions along the LHC is assumed to be given, cf. the example of Eq. (14.20), and it is denoted in the following by $\Delta_L F_i(s)$. Namely,

$$F_i(s + i\varepsilon) - F_i(s - i\varepsilon) = \Delta_L F_i(s) , \ s < s_L . \qquad (15.3)$$

We already have shown in Eq. (13.14) that the $n \times 1$ column vector $F(s)$ of matrix elements $F_i(s)$ can be expressed as the product of the inverse of the $n \times n$ matrix $D(s)$, whose matrix elements are the functions $D_{ij}(s)$, times $L(s)$. The latter is an $n \times 1$ vector column of n analytical functions in the cut complex s plane, $L_i(s)$, which do not have RHC. They could have only LHC (if any) [the possible bound-state poles in $F(s)$ would correspond to zeroes in the $\det D(s)$].

To characterize the different solutions for $F(s)$ it is convenient to introduce an $n \times n$ matrix $S(s)$ defined as

© The Author(s), under exclusive licence to Springer Nature Switzerland AG 2019
J. A. Oller, *A Brief Introduction to Dispersion Relations*, SpringerBriefs in Physics,
https://doi.org/10.1007/978-3-030-13582-9_15

$$S(s) = I + T(s)2i\rho(s) \ . \tag{15.4}$$

Notice that although $T(s)$ is a symmetric this is not the case in general for $\mathcal{S}(s)$. From Eq. (15.4) and the unitarity relation for $T(s)$, Eq. (2.50), it is straightforward to prove that for $s > s_{\text{th};n}$ this matrix satisfies the property,

$$\mathcal{S}(s)\mathcal{S}(s)^* = \mathcal{S}(s)^*\mathcal{S}(s) = I \ . \tag{15.5}$$

To avoid any confusion let us indicate that the asterisk refers to complex conjugation and not to the Hermitian conjugate of the matrix $\mathcal{S}(s)$. Since $\mathcal{S}(s)$ is not symmetric they are not equivalent. Note also the while the S matrix in partial waves, defined in Eq. (2.52), is symmetric and unitary neither of these properties hold in general for $\mathcal{S}(s)$ when $n > 1$.

If we use the N/D method to express $T(s) = D(s)^{-1}N(s)$, we notice that $\mathcal{S}(s)$ can also be written as

$$\begin{aligned}
\mathcal{S}(s) &= I + 2i\,D(s)^{-1}N(s)\rho(s) = D(s)^{-1}\left[D(s) + 2i\,N(s)\rho(s)\right] \\
&= D(s)^{-1}D(s)^* \ ,
\end{aligned} \tag{15.6}$$

an expression valid in the whole complex s plane.

Now, let us assume that we have found an $n \times n$ matrix $D(s)$ with only RHC that satisfies Eq. (15.6). From the previous equation, and taking into account that $D(s)^* = D(s^*)$ it also follows the discontinuity relation

$$D(s)^{-1} = \mathcal{S}(s)D(s^*)^{-1} \ . \tag{15.7}$$

Multiplying both sides by $L(s)$, cf. Eq. (13.14), and taking into account that $L(s)$ is real along the RHC, we then have an analogous relation for the form factors

$$F(s) = \mathcal{S}(s)F(s^*) \ . \tag{15.8}$$

As stated, $L(s) = D(s)F(s)$ has only LHC and its discontinuity along this cut is given by

$$\Delta_L L(s) = D(s)\Delta_L F(s) \ , \tag{15.9}$$

since $D(s)$ is regular along the LHC because of extended unitarity. Assuming that $L(s)$ diverges for $s \to \infty$ less strongly than s^m for some integer $m \geq 0$,[1] we can write the following m-times subtracted DR

---

[1] More rigorously we should say that $L(s)$ diverges less strong than s^{m-1}, $m \geq 1$, to avoid just a logarithmic vanishing of $L(s)/s^m$. However, for the statement above we always have in mind a power-like vanishing, $|L(s)/s^m| < |s|^{-\gamma}$, $\gamma > 0$, for $s \to \infty$.

$$L(s) = \sum_{i=0}^{m-1} a_i s^i + \frac{s^m}{\pi} \int_{-\infty}^{s_L} \frac{D(s')\Delta_L F(s')ds'}{(s')^m(s'-s)} , \tag{15.10}$$

such that if $m = 0$ there is no subtractive polynomial. The latter equation allows to write the following DR representation for $F(s)$,

$$F(s) = D(s)^{-1} \sum_{i=0}^{m-1} a_i s^i + \frac{s^m}{\pi} \int_{-\infty}^{s_L} \frac{D(s)^{-1}D(s')\Delta_L F(s')ds'}{(s')^m(s'-s)} . \tag{15.11}$$

Now, for a given PWA $T(s)$ we can work out $\mathcal{S}(s)$, Eq. (15.5). The problem of finding an $n \times n$ matrix $D(s)$ of functions $D_{ij}(s)$ with only RHC that allows one to write $\mathcal{S}(s) = D(s)^{-1}D(s)^*$ as in Eq. (15.6) is called the Hilbert problem. From Eq. (15.7) it is clear that each column of $D(s)^{-1}$ satisfies the same discontinuity relation as Eq. (15.8) for the form factors along the RHC. Therefore, every column of $D(s)^{-1}$ is itself a form factor with RHC only. The final form factors $F_i(s)$ are obtained by a linear combination of the columns $D(s)^{-1}$, where the coefficients in this linear superposition are the $L_i(s)$ functions that comprise the possible LHC, cf. Eq. (15.10).

First, let us notice that the determinant of the S matrix, $S(s)$, and that of $\mathcal{S}(s)$ are the same,

$$\det S(s) = \det \mathcal{S}(s) . \tag{15.12}$$

This is clear if we consider that

$$\det S = \det \left(I + 2i\rho^{\frac{1}{2}} T \rho^{\frac{1}{2}} \right) = \det \left(\rho^{\frac{1}{2}} \left[\rho^{-\frac{1}{2}} + 2i T \rho^{\frac{1}{2}} \right] \right) = \det \left(I + 2i T \rho \right) = \det \mathcal{S} . \tag{15.13}$$

Notice also that $\det S$ is given by the sum of the eigen-phase shifts $\varphi_i(s)$ as

$$\det S = \exp 2i \sum_{i=1}^{n} \varphi_i(s) = \det \mathcal{S} . \tag{15.14}$$

For the important two-coupled channel case the sum of the eigen-phase shifts is the sum of the phase shifts, as it is clear from Eq. (13.15).

The fact that $\mathcal{S}(s) = D(s)^{-1}D(s)^*$, Eq. (15.6), allows us to write an Omnès representation for $\det D^{-1}$. The point is that $\det D^{-1}$ has only RHC (as the function D itself) and from Eq. (15.6) we learn that the phase of $\det D^{-1}$ is half the phase of $\det S$, which is denoted in the following as $\Phi(s)$, $\Phi(s) = 2 \sum_i \varphi_i(s)$.[2] In addition $\det D^{-1}$ could have zeroes and poles (the former are the generalization of the CDD poles to the coupled-channel case). Out of the zeros and poles of $\det D(s)^{-1}$ we make up the

---

[2] The number of open channels changes. However, $\Phi(s)$ is a continuous function of s along the RHC.

polynomials $P(s)$ and $Q(s)$, respectively, cf. Eq. (14.5). To simplify the notation we further introduce the symbols

$$\Delta(s) = \det D(s)^{-1} , \tag{15.15}$$

$$s_R = s_{\text{th};1} .$$

We take the function $e^{-i\frac{\Phi(s_R)}{2}} Q(s)\Delta(s)/P(s)$, which is amenable to an Omnès representation in the form,

$$\Delta(s) = \frac{P(s)}{Q(s)} \exp \omega(s) , \tag{15.16}$$

$$\omega(s) = \frac{\Phi(s_R)}{2} + \frac{s - s_R}{2\pi} \int_{s_R}^{\infty} \frac{\Phi(s') - \Phi(s_R)}{(s' - s_R)(s' - s)} ds' . \tag{15.17}$$

We have multiplied $\Delta(s)$ by $\exp(-i\Phi(s_R)/2)$ so that the resulting function has a zero phase at s_R, which allows the integral in the DR of the previous equation to stay finite (even if $\Phi(s_R)$ is not zero).

Taking into account that the asymptotic behavior of an Omnès function for $s \to \infty$ is given by the asymptotic phase $\Phi(\infty)$, cf. Eq. (14.10), we then have from Eqs. (15.16) and (15.17) the following limit behavior for $\Delta(s)$,[3]

$$\Delta(s) \xrightarrow[s\to\infty]{} s^{p-q-\frac{\Phi(\infty)-\Phi(s_R)}{2\pi}} . \tag{15.18}$$

Which is the relativistic coupled-channel version of the Levinson theorem, cf. footnote 3.

An interesting result in connection with Eq. (15.18) is that it relates the asymptotic behavior of $\Delta(s)$ with the leading power behavior in s of the columns in $D(s)^{-1}$ [75, 91, 92]. Let $\phi_i(s)$ be the ith column of $D(s)^{-1}$ which, as follows from Eq. (15.7), satisfies the same discontinuity linear relation as a form factor,

$$S(s)\phi_i(s)^* = \phi_i(s) , \quad s > s_R . \tag{15.19}$$

Assuming as in Refs. [75, 92] that $S(s) \to I$ for $s \to \infty$ it is clear that the leading behavior of $\phi_i(s)$ should be integer-power like (no cut remains in this limit and we always assume that all these functions are amenable to a DR treatment). Furthermore, by appropriate linear combinations we can always choose these $\phi_i(s)$ such that if χ_i is the leading degree in s of $\phi_i(s)$ [which corresponds to the degree in s of the dominant component among all the components of $\phi_i(s)$] then

$$\Delta(s) \xrightarrow[s\to\infty]{} s^{\chi_1+\chi_2+\cdots+\chi_n} . \tag{15.20}$$

---

[3] We assume that the zeroes and poles of $\Delta(s)$ do not occur at the threshold s_R.

To see this result let us discuss first the two-coupled channel case and to fix ideas let us assume that $\chi_2 \geq \chi_1$. If the leading behavior for ϕ_2 gives rise to a column vector linearly independent to the leading one for ϕ_1, then the result of Eq. (15.20) is clear. However, if the leading-components vector ϕ_2 is linearly dependent with the leading one from ϕ_1, then multiply ϕ_1 by $s^{\chi_2-\chi_1}$ times a constant and remove it to ϕ_2, which is then the new ϕ_2. In this way (iterated if needed), the leading behavior for $\phi_2(s)$ is now a linearly independent vector to ϕ_1. On the other hand, if $\chi_1 > \chi_2$ we would proceed analogously exchanging $1 \leftrightarrow 2$. It is clear that this process can be further iterated to treat the case with n coupled PWAs, and then Eq. (15.20) results for an appropriately built matrix $D(s)^{-1}$. Notice also that the exponent in the rhs of Eq. (15.20) must match with the one in Eq. (15.18). Thus, we also have that

$$\chi_1 + \chi_2 + \ldots + \chi_n = p - q - \frac{\Phi(\infty) - \Phi(s_R)}{2\pi} \, . \tag{15.21}$$

These results were applied in Ref. [76] to study the strangeness-changing scalar form factors for $K\pi(1)$, $K\eta(2)$ and $K\eta'(3)$, following an analogous set up as in Ref. [83] for the calculation of the $\pi\pi$ and $K\bar{K}$ isoscalar scalar form factors (this latter problem was addressed also by Ref. [84] with a similar approach). The strangeness-changing or $\Delta S = 1$ scalar form factors are defined by

$$\langle 0|\partial^\mu(\bar{s}\gamma_\mu u)(0)|K\phi_K\rangle = -i\sqrt{\frac{3}{2}}\Delta_{K\pi} F_k(s) \, , \tag{15.22}$$

$$\Delta_{K\pi} = m_K^2 - m_\pi^2 \, .$$

The state $|K\pi\rangle$ is in the isospin basis so that its form factor is $\sqrt{3}$ that of $|K^+\pi^0\rangle$, and $|0\rangle$ is the vacuum state.

The $I = 1/2$ scalar $K\pi$, $K\eta'$ PWAs of Ref. [93] were used for driving the FSI. Reference [76] also checked that the results barely change when considering the $K\eta$ channel as well, so that we disregard it in the following and concentrate in the two-coupled channel problem of $K\pi$ and $K\eta'$ scattering. It was further taken for granted in Ref. [76] that the scalar form factors vanish for $s \to \infty$ because the hadrons are composite objects. This is also in agreement with expectations from QCD counting rules [77–79]. As a result, the following unsubtracted DRs were written for the hadronic form factors $F_1(s)$ and $F_3(s)$,

$$F_1(s) = \frac{1}{\pi}\int_{s_{th};1}^\infty \frac{\rho_1(s')F_1(s')T_{11}(s')^* ds'}{s' - s} + \frac{1}{\pi}\int_{s_{th};3}^\infty \frac{\rho_3(s')F_3(s')T_{13}(s')^* ds'}{s' - s} \, ,$$
$$\tag{15.23}$$
$$F_3(s) = \frac{1}{\pi}\int_{s_{th};1}^\infty \frac{\rho_1(s')F_1(s')T_{13}(s')^* ds'}{s' - s} + \frac{1}{\pi}\int_{s_{th};3}^\infty \frac{\rho_3(s')F_3(s')T_{33}(s')^* ds'}{s' - s} \, .$$

These coupled linear IEs were solved numerically in Ref. [93] by iteration. The numerical iterative method developed in this reference is summarized in the appendix.

The PWAs considered in Ref. [93] have no bound states, $q = 0$, and the D^{-1} matrix has no CDD poles [they were reabsorbed in the function $N(s)$], $p = 0$. Furthermore, $\Phi(s_R) = 0$. As a result, the rhs of Eq. (15.18) reads

$$\Delta(s) \xrightarrow[s \to \infty]{} s^{-\frac{\Phi(\infty)}{2\pi}} . \tag{15.24}$$

The first set of T matrices used in Ref. [93], and derived in Ref. [76], give rise to $\Phi(\infty) = 2\pi$ ($\delta_1(\infty) = \pi$ and $\delta_3(\infty) = 0$). It follows then from Eq. (15.21) that

$$\chi_1 + \chi_2 = -1 . \tag{15.25}$$

Since it is not possible that simultaneously χ_1 and χ_2 are negative integers, we then conclude that there is only one linearly independent solution that vanishes at infinity with $\chi_1 = -1$. This is the solution obtained by solving numerically Eq. (15.23) employing the PWAs from the fits (6.10) and (6.11) of Ref. [76]. As starting input Ref. [93] takes for $F_1(s)$ its solution according to an Omnès representation, cf. Eq. (14.5), with constant $P(s)$ and $Q(s)$, while $F_3(s)$ is taking zero initially. The normalization factor corresponds to $F_{K\pi}(0)$ according to its value calculated at NLO in ChPT [94].

Next, Ref. [93] also matched smoothly the unitarized ChPT PWAs of Ref. [76] with a K-matrix ansatz at an energy around $\sqrt{s} = 1.75$ GeV. The point is to improve the reproduction of the experimental data of Ref. [95] on $K\pi$ scattering, that was somewhat deficiently accomplished by the PWAs of Ref. [76] for energies above 1.9 GeV. Once this is done Ref. [93] could consider the transition from $\Phi(\infty) = 2\pi$ to $\Phi(\infty) = 4\pi$ by changing some suitable parameters in the K-matrices employed, while reproducing satisfactorily the experimental data up to the largest energy available in Ref. [95] ($\sqrt{s} = 2.5$ GeV). For the case $\Phi(\infty) = 4\pi$ we then have from Eq. (15.21) that ($q = p = \Phi(s_R) = 0$)

$$\chi_1 + \chi_2 = -2 . \tag{15.26}$$

In this case we can then have two linearly independent solutions with negative χ_i for $\chi_1 = \chi_2 = -1$. This second linearly independent solution was found in Ref. [93] by solving Eq. (15.23) with different input values for the form factors at the origin. Apart from a global normalization another piece of information is needed, since now there are two linearly independent solutions. The Ref. [93] uses the value of the $K\pi$ form factor at the Callan-Treiman point, where $s = \Delta_{K\pi}$, because it can be related quite accurately with the ratio of the weak decay constants of the pseudoscalar kaons (f_K) and pions (f_π). The precise relation is

$$F_{K\pi}(\Delta_{K\pi}) = \frac{f_K}{f_\pi} + \Delta_{CT} , \tag{15.27}$$

with Δ_{CT} estimated as -3×10^{-3} [94], while f_K/f_π is taken in Ref. [93] as 1.22 ± 0.01, according to the phenomenological information then available. It is worth emphasizing that once the chiral unitary amplitudes of Ref. [76] were implemented with the K-matrix ansätze, independently of whether $\Phi(\infty) = 2\pi$ (one linearly independent solution) or 4π (two linearly independent solutions) the value of $F_{K\pi}(\Delta_{K\pi})$ is in both cases compatible, which indicates the great stability of the results. For the case with only one linearly independent solution it turns out that $F_{K\pi}(\Delta_{K\pi}) = 1.219 - 1.22$ in impressive agreement with Eq. (15.27).

Let us finish this section by connecting with the use of the function $\mathcal{N}(s)$ to express the form factors $F_i(s)$ as in Eq. (13.11), by using the matrix of functions $[I + \mathcal{N}(s)g(s)]^{-1}$. In the special case in which $\mathcal{N}(s)$ is modeled without LHCs, as discussed in Chap. 6, we could end with explicit expressions for $\Omega(s)$ (in the uncoupled case) and for $D(s)$ in the coupled case. For the former case we would have

$$\Omega(s) = \frac{\prod_{i=1}^{q}(s - s_{P;i})}{\prod_{j=1}^{p}(s - s_{Z;j})} \frac{1}{1 + \mathcal{N}(s)g(s)} , \tag{15.28}$$

with the subscripts P and Z referring to the poles and zeroes of $1/[1 + \mathcal{N}(s)g(s)]$, which are removed by multiplying this function by the appropriate rational function. For the case of coupled channels, we can identify the matrix $D(s)$ in Eq. (15.7) with

$$D(s) = [I + \mathcal{N}(s)g(s)] . \tag{15.29}$$

We can also introduce like in Ref. [92] the analogous of $\Omega(s)$ in coupled channels, denoted by $\mathcal{D}^{-1}(s)$, so that $\mathcal{D}^{-1}(s)$ satisfies Eq. (15.7), it is holomorphic and nonsingular in the cut complex s plane. Given the function $D(s)$ in Eq. (15.29) we notice that the product $\mathcal{D}(s)D(s)^{-1}$ has no cuts because from Eq. (15.7)

$$\mathcal{D}(s + i\varepsilon)D(s + i\varepsilon)^{-1} - \mathcal{D}(s - i\varepsilon)D(s - i\varepsilon)^{-1} \tag{15.30}$$

$$= \mathcal{D}(s - i\varepsilon)\mathcal{S}(s - i\varepsilon)\mathcal{S}(s + i\varepsilon)D(s - i\varepsilon)^{-1} - \mathcal{D}(s - i\varepsilon)D(s - i\varepsilon)^{-1} = 0 ,$$

taking also into account Eq. (15.5) and that $\mathcal{D}(s^*) = \mathcal{D}(s)^*$. The same property would also hold for $D(s)\mathcal{D}(s)^{-1}$. Therefore, the product $\mathcal{D}(s)D(s)^{-1}$ is a rational function $R(s)$ and we can write [92]

$$D(s)^{-1} = \mathcal{D}(s)^{-1}R(s) . \tag{15.31}$$

Of course, this result applies to any possible matrix of functions $D(s)$ satisfying Eq. (15.6), independently of the modeling of $\mathcal{N}(s)$.

Chapter 16
Near-Threshold Scattering

In this section we consider the physics in a near-threshold region so that the nonrelativistic limit is appropriate. Furthermore, we assume that the LHC is relatively weak and/or far away. In the later case the LHC admits a Taylor expansion in the region of interest and its effects can be accounted for without explicitly including it. For the former case we assume that it can be neglected in good approximation because of its weakness. With this scenario in mind we use the general results deduced in Chap. 6, cf. Eq. (6.24), which are applicable when the cut associated with the LHC is not explicitly realized.

We consider the S-wave scattering, which is expected to be dominant since the energy is supposed to be near the threshold of the reaction of the two particles with masses m_1 and m_2. In such circumstances the general structure of a PWA corresponds to Eq. (6.24) with $L = 0$. The relativistic phase space in the integral along the RHC is $p(s)/8\pi\sqrt{s}$. We also introduce the kinetic energy E by its nonrelativistic expression, namely,

$$E = \frac{p^2}{2\mu} \, , \tag{16.1}$$

$$p = \sqrt{2\mu E} \, ,$$

which is more appropriate for nonrelativistic dynamics, with the relation $\sqrt{s} = m_1 + m_2 + E + \mathcal{O}(p^4)$.

The unitary loop function, that corresponds to the integral along the RHC in Eq. (6.24), is already given in Eq. (8.3). Its series in powers of $p = |\mathbf{p}|$ around threshold gives rise to an expansion involving powers of p^2 [it corresponds to the expansion in real variable of its real part for physical values of $s > s_{\text{th}}$. This is given by the Cauchy's principal value of the integral in Eq. (8.3)] and odd powers of p [which stems from the expansion of its imaginary part $-p/8\pi\sqrt{s}$]. The first terms of this nonrelativistic series of $g(p)$ in powers of p read

© The Author(s), under exclusive licence to Springer Nature Switzerland AG 2019
J. A. Oller, *A Brief Introduction to Dispersion Relations*, SpringerBriefs in Physics,
https://doi.org/10.1007/978-3-030-13582-9_16

$$g(p) = \frac{a}{16\pi^2} + \frac{1}{8\pi^2(m_1 + m_2)}(m_1 \log \frac{m_1}{\mu} + m_2 \log \frac{m_2}{\mu}) \qquad (16.2)$$

$$- i \frac{p}{8\pi(m_1 + m_2)} + \mathcal{O}(\frac{p^2}{\Sigma^2}) ,$$

where Σ has dimension of mass and it is made out of the masses m_1 and m_2. For every $\mathcal{O}(p^m)$ term with $m \geq 1$ there is always a neat power of mass m_i, $i = 1, 2$, in the denominator, avoiding any relative enhancement from powers of the factor m_1/m_2 with $m_1 \gg m_2$. This is clear from the nonrelativistic reduction of the RHC integral in Eq. (8.3),

$$g(p) = \tilde{a} - \frac{s}{\pi} \int_0^\infty \frac{dq^2}{\omega_1'(q)\omega_2'(q)} \frac{q}{8\pi[\omega_1'(q) + \omega_2'(q)]} \qquad (16.3)$$

$$\times \frac{1}{[\omega_1'(q) + \omega_2'(q) - \omega_1(p) - \omega_2(p)][\omega_1'(q) + \omega_2'(q) + \omega_1(p) + \omega_2(p)]}$$

$$= \tilde{a} + \frac{1}{8\pi(m_1 + m_2)} \int_0^\infty \frac{qdq^2}{p^2 - q^2} + \mathcal{O}(\frac{p^2}{\Sigma^2}) ,$$

with $\omega_i(q)$ already introduced in Eq. (8.5).

In the following we adopt the more standard nonrelativistic normalization of the PWA $t(p^2)$, already introduced in Eq. (11.9) for $m_1 = m_2 = m$. Additionally, we denote by β the momentum-independent contribution on the rhs of Eq. (16.2) (times $8\pi(m_1 + m_2)$ because of the change in normalization). Namely,

$$\beta = \frac{a(m_1 + m_2)}{2\pi} + \frac{1}{\pi}(m_1 \log \frac{m_1}{\mu} + m_2 \log \frac{m_2}{\mu}) . \qquad (16.4)$$

Attending to Eq. (6.24) with $L = 0$, in addition to the subtraction constant we also have the sum over the CDD poles. Looking for relevant structures in the near-threshold region apart from the threshold branch-point singularity, we explore the consequences of including a CDD pole. In this way, we recast Eq. (6.24) as

$$t(E) = \left(\frac{\gamma}{E - M_{\text{CDD}}} + \beta - ip(E) \right)^{-1} , \qquad (16.5)$$

where we use the kinetic energy E as variable, cf. Eq. (16.1). In this equation γ is the residue of the CDD pole and M_{CDD} is its position in energy E.

Despite the straightforward derivation of Eq. (16.5) by attending to basic analytical properties of PWAs, in this case, the presence of the RHC and of a pole in the inverse of the PWA, goes beyond an ERE, up to an including $\mathcal{O}(p^4)$, cf. Eq. (11.10). The reason is because the presence of a zero in $t(p^2)$ [or a pole in $1/t(p^2)$] sets a limit in the applicability of the ERE because at this point $p \cot \delta = \infty$ and it is singular. Thus, if this zero happens very close to threshold it makes the ERE to have a very small region of validity, which would typically invalidate it as an adequate approach to study the near-threshold scattering. One can work out straightforwardly the relationship between the parameters a, r_2, and v_2 in the ERE with γ, M_{CDD} and β in Eq. (16.5), it reads

$$\frac{1}{a} = \frac{\gamma}{M_{CDD}} - \beta \,, \tag{16.6}$$

$$r = -\frac{\gamma}{\mu M_{CDD}^2} \,,$$

$$v_2 = -\frac{\gamma}{4\mu^2 M_{CDD}^3} \,.$$

An important output of these relations [96] is that a near-threshold CDD pole, $M_{CDD} \to 0$, is characterized by giving rise to large values of r in absolute module (γ can take any sign). This is also the expected situation for v_2 and higher shape parameters v_i, $i \geq 4$. However, the value for the scattering length in the same limit $M_{CDD} \to 0$ would tend to zero as M_{CDD}/γ. Of course, the actual situation in which this limit takes place depends on the value of the residue of the CDD pole, the larger is γ the sooner this scenario takes place.

Another parameterization that is usually employed in the literature to describe near-threshold resonances is the so-called Flatté parameterization [97], that we denote as $t_F(E)$ and corresponds to

$$t_F(E) = \frac{g^2/2}{M_F - i\frac{1}{2}\Gamma(E) - E} \,, \tag{16.7}$$

$$\Gamma(E) = g^2 p(E) \,, E > 0 \,,$$

$$\Gamma(E) = ig^2|p(E)| \,, E < 0 \,,$$

and $\Gamma(E) \geq 0$ for $E > 0$, which determines that $g^2 \geq 0$. The Flatté mass M_F is the value of the energy for which the real part of the denominator in $t_F(E)$ vanishes. The energy dependence of $\Gamma(E)$ is a characteristic aspect of a Flatté parameterization.

We notice here that $t_F(E)$ is a particular case of an ERE, with the denominator in Eq. (16.7) involving up to quadratic powers in p. The relationship between the ERE parameters a, r and those in the Flatté parameterization g^2, M_F is

$$a = -\frac{g^2}{2M_F} \,, \tag{16.8}$$

$$r = -\frac{2}{g^2\mu} \,.$$

Thus, a Flatté parameterization can only give rise to negative values for the effective range, $r < 0$. The scattering length changes of sign with M_F and, for fixed coupling g^2, it is infinity for $M_F = 0$, in which case $t_F(0)$ becomes infinity too. There is indeed a qualitatively different behavior of the pole content of $t_F(E)$ depending on whether M_F is positive or negative. Solving the roots in p of the denominator of $t_F(E)$, we can write the latter as

$$t_F(E) = \frac{-\mu g^2}{(p(E) - p_1)(p(E) - p_2)} , \qquad (16.9)$$

$$p_{1,2} = -i\frac{g^2\mu}{2}\left(1 \pm \sqrt{1 - \frac{8M_F}{g^4\mu}}\right) , \qquad (16.10)$$

with the subscript 1(2) corresponding to the +(−) sign in front of the square root. Thus, if $M_F \leq 0$ then $p_{1,2}$ are purely imaginary, but with opposite signs. In this way, $\Im p_1 < 0$ and it corresponds to a virtual state in the second RS, while $\Im p_2 > 0$ and it gives rise to a bound state (in the first RS). Furthermore, $|p_1| > |p_2|$ and the virtual state is deeper than the bound state, which is then closer to threshold. For $M_F = 0$ the bound state has zero binding energy. When M_F becomes positive and lies in the interval $0 < M_F < g^4\mu/8$, the second pole turns out as another virtual state closer to threshold than p_1.

For $M_F > g^4\mu/8$ the two pole positions $p_{1,2}$ have the same negative imaginary part but real parts with opposite signs. These are poles corresponding to resonances, such that in the limit $M_F \to g^4\mu/8 + \varepsilon$ the real part tends to zero and we would end with a a double virtual-state pole [98]. This is also a limitation of the Flatté model, no higher than double poles can arise from this parameterization.

The resonance poles happens in complex conjugate positions in the complex E plane ($E = p^2/2\mu$), which is generally required because of the fulfillment by the PWAs of the Schwarz reflection principle. For this situation, we read from Eq. (16.7) that in the limit in which $\Gamma(M_F) \ll M_F$ the nearest pole position to the physical axis occurs in good approximation at

$$E_F = M_F - i\frac{\Gamma(M_F)}{2} , \qquad (16.11)$$

where the equation $M_F - i\Gamma(E)/2 - E = 0$ is solved by iterating it once in $\Gamma(E)$. This is the situation corresponding to the narrow resonance case. The other pole at $E = E_F^*$ in the second RS is further away from the physical or first RS, because the physical values are obtained in the latter sheet by taking $E + i\varepsilon$ with $\varepsilon \to 0^+$. This upper part of the physical axis is connected smoothly with the negative vanishing imaginary part in the second RS, which is the region in which the approximate pole position of Eq. (16.11) lies. Nonetheless, the pole with positive complex imaginary part is connected with the values of the scattering amplitudes in the complex E plane below the real axis (where the scattering amplitude is the complex conjugate of the physical one).

The poles of $t_F(E)$ in the second RS of the complex E plane are $E_{1,2} = p_{1,2}^2/2\mu$, with $p_{1,2}$ given in Eq. (16.10). The corresponding expressions for $E_{1,2}$ are

$$E_{1,2} = M_F - \frac{g^4\mu}{4} \mp i\frac{g^4\mu}{4}\sqrt{\frac{8M_F}{g^4\mu} - 1} \, , \ M_F > \frac{g^4\mu}{8} \, , \tag{16.12}$$

There is a variation of the real part of $E_{1,2}$ with respect to M_F due to the self-energy contribution $-g^4\mu/4$. On the one hand, from Eq. (16.12) we identify the resonance mass, M_R, as $\Re E_{1,2}$,

$$M_R = M_F - \frac{g^4\mu}{4} \, . \tag{16.13}$$

On the other hand, twice the modulus of the imaginary part of $E_{1,2}$ is identified with the width of the resonance, Γ, given then by

$$\Gamma = \frac{g^4\mu}{2}\sqrt{\frac{8M_F}{g^4\mu} - 1} \, . \tag{16.14}$$

Let us notice that, this expression for Γ is only equal to $g^2\sqrt{2\mu M_R}$, cf. Eq. (16.7), for the case in which $M_F \gg g^4\mu$. In this situation the expression for Γ in Eq. (16.11) also holds in good approximation. Thus, the narrow resonance limit actually requires that $M_F \gg g^4\mu$.

It is also interesting to workout the residues of $t_F(E)$ at the pole positions, either in the complex momentum or energy spaces,

$$\gamma_k^2 = -\lim_{p \to p_i} (p - p_i)t_F(p^2/2\mu) \, , \tag{16.15}$$

$$\gamma_E^2 = -\lim_{E \to E_p} (E - E_i)t_F(E) \, , \tag{16.16}$$

in order. Both types of residues are related by

$$\gamma_E^2 = \gamma_k^2 \left.\frac{dE}{dp}\right|_{p_{1,2}} = g_k^2 \frac{p_{1,2}}{\mu} \, . \tag{16.17}$$

Working out the residue γ_k^2 is straightforward from Eq. (16.9), with the result

$$\gamma_k^2 = \pm\frac{\mu g^2}{p_1 - p_2} = \pm\frac{1}{\sqrt{\frac{8M_F}{g^4\mu} - 1}} \, , \tag{16.18}$$

with $+(-)$ for the pole $p_1(p_2)$, in this order. Notice that in the narrow resonance case, $M_F \gg g^4\mu/8$, the coupling $\gamma_k^2 \to 0$. The opposite situation occurs for $M_F \to g^4\mu/8$ in which case the coupling diverges, because we end with a double virtual-state pole [98]. Let us recall that this is the starting point for having resonance poles, cf. Eq. (16.12).

There is also another interesting limit that corresponds to the case in which the real part of $E_{1,2}$ starts becoming positive. According to Eq. (16.12), this occurs for $g^4\mu/4 < M_F < \infty$ and then $1 \geq |\gamma_k|^2 \geq 0$. Indeed, one can develop a probabilistic interpretation for $|\gamma_k|^2$ when the real part of the pole position in energy of the resonance is larger than zero. According to this interpretation, $|\gamma_k|^2$ is the weight of the two-body continuum component in the composition of the resonance [96, 99].

As noted above, the effective range r for a Flatté parameterization must be negative, cf. Eq. (16.8). In general terms, a PWA $t(p^2)$ from an ERE up to and including p^2, cf. Eq. (11.9), and denoted by $t_r(E)$, is given by

$$t_r(E) = \frac{1}{-\frac{1}{a} + \frac{1}{2}rp(E)^2 - ip(E)} \, . \tag{16.19}$$

Given the quadratic nature of the denominator in p we also have two poles corresponding to the values

$$p_{1,2} = \frac{1}{r}\left(i \mp \sqrt{\frac{2r}{a} - 1}\right) \, . \tag{16.20}$$

We have resonance poles for

$$r/a > 1/2 \text{ and } r < 0 \, . \tag{16.21}$$

Notice that the imaginary part for a resonance pole should be negative as it lies on the second RS, and this is why we have required that $r < 0$. Let us also indicate that, if the requirements in Eq. (16.21) are applied to the expressions for a and r of a Flatté parameterization in Eq. (16.8), we have the constraint $M_F > g^2\mu/8$, which we already derived as necessary so as to end with resonance poles. It is also interesting to work out the residues γ_k^2 of $t_r(E)$ for the poles in Eq. (16.20). The corresponding expression is

$$\gamma_k^2 = \frac{1}{rp_{1,2} - i} = \mp\frac{1}{\sqrt{\frac{2r}{a} - 1}} \, . \tag{16.22}$$

The requirement for $0 \leq |\gamma_k^2| \leq 1$ implies that

$$\frac{r}{a} \geq 1 \, . \tag{16.23}$$

Again, if we consider this constraint in terms of the values of a and r as corresponding to the Flatté parameterization we then have the condition $M_F \geq g^2\mu/4$, which is also needed so that the real part of the resonance energy is positive. Indeed, from Eq. (16.20) the pole energy $E_{1,2} = p_{1,2}^2/2\mu$ is given by

$$E_{1,2} = \frac{p_{1,2}^2}{2\mu} = \frac{1}{ar\mu}\left(1 - \frac{a}{r}\right) \mp i\frac{1}{r^2\mu}\sqrt{\frac{2r}{a} - 1} \,. \tag{16.24}$$

It is then concluded form the previous expression that $r/a > 1$ so that $\Re E_{1,2} \geq 0$, and then the criterion for the probabilistic interpretation of $|\gamma_k|^2$, as developed in Refs. [96, 99], can be applied. From Eq. (16.24) we also have for the width of the resonance (identified as twice the modulus of the imaginary part of $E_{1,2}$),

$$\Gamma = \frac{2}{r^2\mu}\sqrt{\frac{2r}{a} - 1} \,. \tag{16.25}$$

From Eq. (16.22) $|\gamma_k|^2$ decreases as r/a increases. At this point it is worth connecting with the values of a and r given by a near-threshold CDD pole, worked out in Eq. (16.6). From this equation it follows that as $M_{CDD} \to 0$ one is driven towards values of a and r for which $|r/a|$ increases and, therefore, $|\gamma_k|^2$ decreases. This fact is interpreted as that the resonance when $M_{CDD} \to 0$ becomes purely elementary, in the sense that the weight in the resonance state of the two-body asymptotic states, whose scattering is described by $t_r(E)$, tends to vanish. Precisely, this is connected with the standard interpretation for a CDD pole which is typically associated with the need to introduce explicitly in the equations the exchange of an explicit bare resonance. In particular, notice that a bare resonance is characterized by two basic parameters, its mass and coupling to a given state. Similarly, a CDD pole implies two free parameters, its mass and residue. According to Eq. (16.24) in the limit $M_{CDD} \to 0$ we have for the resonance poles positions

$$E_{1,2} \xrightarrow[M_{CDD}\to 0]{} -\frac{M_{CDD}^3}{\lambda^2} \mp i\frac{(-M_{CDD})^{7/2}\sqrt{2\mu}}{\lambda^2} \tag{16.26}$$

and we see that the width vanishes faster than the mass (the real part of $E_{1,2}$) by an extra factor $(-M_{CDD})^{1/2}$. This pole is then characterized by a small mass but even a much smaller width, so that the narrow resonance limit holds. Indeed, the decoupling limit of a bare resonance from the two-body continuum requires a zero in the PWA in order to remove the bare pole of the resonance from $t(E)$. This shows in simple terms that the weak coupling limit of a resonance and the presence of a CDD pole are related. We also mention that in order to fulfill the requirements in Eq. (16.21) it is necessary to have negative M_{CDD} and positive γ in the limit $M_{CDD} \to 0$.

The situation in Eq. (16.26) is opposite to the one when $\gamma_k^2 \to 1$, which according to Eq. (16.22) happens for $r/a \to 1$. In such a case, we infer from Eq. (16.24) that the mass of the resonance vanishes in this limit and the energy becomes purely imaginary and finite. Therefore, a resonance that is purely composite of the asymptotic two-body state whose interaction is given by $t_r(E)$ is characterized by having a width much larger than its mass.

We have shown that the parameterization in Eq. (16.5) is more general than an ERE up to an including $\mathcal{O}(p^4)$, because the former accounts for the possibility of near-threshold zeroes while the latter loses its meaning for energies beyond the zero and its convergence is much worse for lower energies. If the parameterization based on the ERE is restricted to terms up to $\mathcal{O}(p^2)$ in the p^2 expansion one then has a PWA, Eq. (16.19), that is more general than the one obtained by applying the Flatté parameterization, Eq. (16.7), because the later can only give rise to negative values of r.

Another type of parameterization that one usually finds in the literature for describing near-threshold scattering stems from the use of a dynamical model based on solving a LS equation with a potential that also includes the exchange of an explicit bare resonance. This is a definite model that exemplifies the connection between the exchange of a bare resonance and the appearance of a CDD pole in the PWA, as commented above. We qualify the scattering due to an energy-independent potential $V(\mathbf{p}, \mathbf{p}')$ as direct scattering between the two-body states in the continuum. On top of it, the exchange of a bare state is also considered, so that the total potential in the continuum, $V_T(\mathbf{p}, \mathbf{p}', E)$, is given by

$$V_T(\mathbf{p}, \mathbf{p}', E) = V(\mathbf{p}, \mathbf{p}') + \frac{f(\mathbf{p}) f(\mathbf{p}')}{E - E_0} . \qquad (16.27)$$

Here E_0 is the bare mass of the discrete state that is exchanged. The real function $f(\mathbf{p})$ is the bare coupling of this state to the two-body states. The scattering amplitude is given by solving the LS equation in momentum space, cf. Eq. (2.65),

$$T(\mathbf{p}, \mathbf{p}', E) = V_T(\mathbf{p}, \mathbf{p}', E) + \int \frac{d^3 q}{(2\pi)^3} \frac{V_T(\mathbf{p}, \mathbf{q}, E) T(\mathbf{q}, \mathbf{p}', E)}{q^2/(2\mu) - E - i\varepsilon} . \qquad (16.28)$$

The solution of this IE is clear and intuitive by employing a graphical method. First consider those diagrams without the exchange of any bare-state propagator. This is represented in the panel (a) of Fig. 16.1, where the point vertices, each with four lines attached, indicate the insertion of a factor of $V(\mathbf{q}, \mathbf{q}')$. In turn, the circles joining vertices correspond to the loops with two propagators associated with the two-body intermediate states in the continuum. The panel (a) of Fig. 16.1 represents the iteration of $V(\mathbf{p}, \mathbf{p}')$ that gives rise to the direct-scattering amplitude $T_V(\mathbf{p}, \mathbf{p}', E)$, that results by solving the LS equation of the pure potential problem,

$$T_V(\mathbf{p}, \mathbf{p}', E) = V(\mathbf{p}, \mathbf{p}') + \int \frac{d^3 q}{(2\pi)^3} \frac{V(\mathbf{p}, \mathbf{q}) T_V(\mathbf{q}, \mathbf{p}', E)}{q^2/(2\mu) - E - i\varepsilon} . \qquad (16.29)$$

Next, we consider those contributions containing at least the exchange of one bare state, which is represented pictorially by a double line. When iterating these contributions we have as intermediate states both two particles in the continuum and extra bare-state exchanges. In this way, we have the standard Dyson resummation for the bare-state propagator, giving rise to the dressed one, as represented in the panel (b) of Fig. 16.1. In addition, we also have the dressing of the bare coupling of the exchanged state to the continuum by the direct scattering of the latter, as represented

(a)

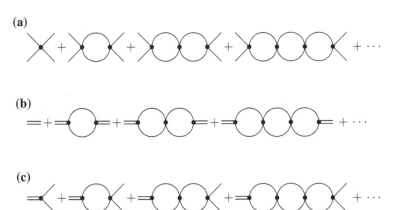

(b)

(c)

Fig. 16.1 Diagrammatic representation of the solution for $T(\mathbf{p}, \mathbf{p}', E)$ in Eq. (16.28). The diagrams in **a** represents the iteration of $V(\mathbf{p}, \mathbf{p}')$ without any bare-state exchange, which generates $T_V(\mathbf{p}, \mathbf{p}', E)$, Eq. (16.29). The panel **b** represents the self-energy for getting the dressed propagator. Those diagrams in panel **c** correspond to the dressing of the bare coupling due to the self-interactions (or final-state interactions) between the two-body states in the continuum

in the panel (c) of Fig. 16.1. Thus, the set of diagrams in the panels (b) and (c) of Fig. 16.1 gives rise finally to the exchange of a particle with dressed propagator and couplings, in the form

$$R(\mathbf{p}, \mathbf{p}', E) = \frac{\Theta(\mathbf{p}, E)\Theta(\mathbf{p}', E)}{E - E_0 + G(E)} , \qquad (16.30)$$

where $\Theta(\mathbf{p}, E)$ represents the dressed coupling and $1/[E - E_0 + G(E)]$ the dressed propagator. We then conclude that the scattering amplitude $T(\mathbf{p}, \mathbf{p}', E)$ must be given by the sum of T_V and R,

$$T(\mathbf{p}. \mathbf{p}', E) = T_V(\mathbf{p}, \mathbf{p}', E) + \frac{\Theta(\mathbf{p}, E)\Theta(\mathbf{p}', E)}{E - E_0 + G(E)} . \qquad (16.31)$$

First, we are going to show directly that indeed Eq. (16.31) is a solution of the LS equation in Eq. (16.28), for appropriate functions $\Theta(\mathbf{p}, E)$ and $G(E)$. Next, we give a more general derivation of the solution for the LS equation in terms of the solution of another LS equation with one less discrete intermediate state. In the present example for the total potential in Eq. (16.27), this is the scattering amplitude $T_V(\mathbf{p}, \mathbf{p}', E)$, which satisfies the LS equation of Eq. (16.29) without the intermediate bare state.

By inserting the tentative solution of Eq. (16.31) into Eq. (16.28), and taking into account that $T_V(\mathbf{p}, \mathbf{p}', E)$ fulfills Eq. (16.29), we are then left with the following IE for $R(\mathbf{p}, \mathbf{p}', E)$,

$$\frac{\Theta(\mathbf{p}, E)\Theta(\mathbf{p}', E)}{E - E_0 + G(E)} = \frac{f(\mathbf{p})f(\mathbf{p}')}{E - E_0} + \int \frac{d^3q}{(2\pi)^3} \frac{1}{q^2/(2\mu) - E - i\varepsilon} \left[V(\mathbf{p}, \mathbf{q}) \frac{\Theta(\mathbf{q}, E)\Theta(\mathbf{p}', E)}{E - E_0 + G(E)} \right.$$

$$\left. + \frac{f(\mathbf{p})f(\mathbf{q})}{E - E_0} T_V(\mathbf{q}, \mathbf{p}', E) + \frac{f(\mathbf{p})f(\mathbf{q})}{E - E_0} \frac{\Theta(\mathbf{q}, E)\Theta(\mathbf{p}', E)}{E - E_0 + G(E)} \right]. \qquad (16.32)$$

We can derive the equations satisfy by $G(E)$ and $\Theta(\mathbf{p}, E)$ by taking $E \to E_0$ and $E \to E_0 - G(E)$ in the previous equation. In order, we are then left with

$$\Theta(\mathbf{p}', E) \frac{-1}{G(E)} \int \frac{d^3q}{(2\pi)^3} \frac{f(\mathbf{q})\Theta(\mathbf{q}, E)}{q^2/(2\mu) - E - i\varepsilon} = f(\mathbf{p}') + \int \frac{d^3q}{(2\pi)^3} \frac{f(\mathbf{q})T_V(\mathbf{q}, \mathbf{p}', E)}{q^2/(2\mu) - E - i\varepsilon}. \qquad (16.33)$$

$$\Theta(\mathbf{p}, E) = -\frac{f(\mathbf{p})}{G(E)} \int \frac{d^3q}{(2\pi)^3} \frac{f(\mathbf{q})\Theta(\mathbf{q}, E)}{q^2/(2\mu) - E - i\varepsilon} + \int \frac{d^3q}{(2\pi)^3} \frac{V(\mathbf{p}, \mathbf{q})\Theta(\mathbf{q}, E)}{q^2/(2\pi) - E - i\varepsilon}.$$

These two equations can be made equivalent by identifying[1]

$$G(E) = -\int \frac{d^3q}{(2\pi)^3} \frac{f(\mathbf{q})\Theta(\mathbf{q}, E)}{q^2/(2\mu) - E - i\varepsilon}, \qquad (16.34)$$

and requiring that $\Theta(\mathbf{p}, E)$ satisfies the inhomogeneous IE

$$\Theta(\mathbf{p}', E) = f(\mathbf{p}') + \int \frac{d^3q}{(2\pi)^3} \frac{f(\mathbf{q})T_V(\mathbf{q}, \mathbf{p}', E)}{q^2/(2\mu) - E - i\varepsilon}. \qquad (16.35)$$

Let us notice that this IE can also be rewritten as

$$\Theta(\mathbf{p}', E) = f(\mathbf{p}') + \int \frac{d^3q}{(2\pi)^3} \frac{T_V(\mathbf{p}', \mathbf{q}, E)f(\mathbf{q})}{q^2/(2\mu) - E - i\varepsilon} \qquad (16.36)$$

$$= f(\mathbf{p}') + \int \frac{d^3q}{(2\pi)^3} \frac{V(\mathbf{p}', \mathbf{q})\Theta(\mathbf{q}, E)}{q^2/(2\mu) - E - i\varepsilon}.$$

The three IEs in Eqs. (16.35) and (16.36) are equivalent as it is clear by performing the Neumann series expansion of $T_V(\mathbf{q}, \mathbf{p}', E)$ from Eq. (16.35), and by solving iteratively the last IE for $\Theta(\mathbf{p}', E)$ in Eq. (16.36). It is straightforward to show that Eq. (16.32) is fulfilled once Eqs. (16.34) and (16.35) are satisfied. For instance, by inserting Eq. (16.35) in Eq. (16.32), we can combine the first and third terms on the rhs of this equation as $f(\mathbf{p})\Theta(\mathbf{p}', E)/(E - E_0)$. Therefore, we can simplify the factor $\Theta(\mathbf{p}', E)$ on both sides of the resulting equation, which then reads

---

[1] In Eq. (16.33) we have renamed E_0 as E because there is nothing special on E_0, so that it can also be considered as a variable energy.

$$\frac{\Theta(\mathbf{p}, E)}{E - E_0 + G(E)} = \frac{f(\mathbf{p})}{E - E_0} + \int \frac{d^3q}{(2\pi)^3} \frac{1}{q^2/(2\mu) - E - i\varepsilon} \left[\frac{V(\mathbf{p}, \mathbf{q})\Theta(\mathbf{q}, E)}{E - E_0 + G(E)} \right.$$

$$\left. + \frac{f(\mathbf{p})f(\mathbf{q})\Theta(\mathbf{q}, E)}{(E - E_0)(E - E_0 + G(E))} \right] \tag{16.37}$$

$$= \frac{f(\mathbf{p})}{E - E_0} + \frac{1}{E - E_0 + G(E)} \int \frac{d^3q}{(2\pi)^3} \frac{V(\mathbf{p}, \mathbf{q})\Theta(\mathbf{q}, E)}{q^2/(2\mu) - E - i\varepsilon} - \frac{f(\mathbf{p})G(E)}{(E - E_0)(E - E_0 + G(E))}$$

$$= \frac{f(\mathbf{p})}{E - E_0 + G(E)} + \int \frac{d^3q}{(2\pi)^3} \frac{1}{q^2/(2\mu) - E - i\varepsilon} \frac{V(\mathbf{p}, \mathbf{q})\Theta(\mathbf{q}, E)}{E - E_0 + G(E)},$$

which is fulfilled in virtue of Eq. (16.36).

Now, let us give a more general derivation of the separation of the total T-matrix $T(\mathbf{p}, \mathbf{p}', E)$ as in Eq. (16.31). Let the full Hamiltonian H be split as in Chap. 2 in the free part H_0 and the potential V, $H = H_0 - V$, and let $|0\rangle$ be an eigenstate of H_0, $H_0|0\rangle = E_0|0\rangle$. Let $T_1(E)$ be the T matrix that fulfills a LS equation without the discrete intermediate state $|0\rangle$, namely,

$$T_1(E) = V + \sum_n \int dW \, V|W_n\rangle(W_n - E)^{-1}\langle W_n|T_1(E) \tag{16.38}$$

$$= V + V(H_0 - E)^{-1}\theta T_1(E),$$

where $\theta|0\rangle = 0$ and this state is then excluded in the sum over intermediate states. In the previous equation we have also used a compressed notation for the sum over discrete states and indexes, represented by n ($n \neq 0$), and integration over the continuum ones, represented by W. The corresponding intermediate state is then indicated by $|W_n\rangle$. The Eq. (16.38) can be recast as the IE for the resolvent of the kernel of a linear IE. For that, we multiply this equation to the right by $(H_0 - E)^{-1}\theta$, which then reads

$$T_1(E)(H_0 - E)^{-1}\theta = V(H_0 - E)^{-1}\theta + V(H_0 - E)^{-1}\theta T_1(E)(H_0 - E)^{-1}\theta. \tag{16.39}$$

The kernel of this IE is $V(H_0 - E)^{-1}\theta$ and its resolvent $K_1(E)$ is therefore [18]

$$K_1(E) = T_1(E)(H_0 - E)^{-1}\theta. \tag{16.40}$$

Now, we take into account that we can formally write from the LS that $T(E) = V(I - (H_0 - E)^{-1}V)^{-1} = V(H - E)^{-1}(H_0 - E)$. It follows then that the resolvent of the kernel of the LS equation, as identified in Eq. (16.40), is $T(E)(H_0 - E)^{-1} = V(H - E)^{-1}$, cf. Eq. (2.64). The inclusion of θ in Eq. (16.40) is just a projection in the subspace orthogonal to $|0\rangle$.

The full T matrix $T(E)$ satisfies a LS equation in which the state $|0\rangle$ contributes as intermediate state. Thus,

$$T(E) = V + V|0\rangle(E_0 - E)^{-1}\langle 0|T(E) + \sum_n \int dW V|W_n\rangle(W_n - E)^{-1}\langle W_n|T(E)$$

$$\tag{16.41}$$

$$= V + V|0\rangle(E_0 - E)^{-1}\langle 0|T(E) + V(H_0 - E)^{-1}\theta T(E) .$$

We can clearly identify from the previous equation the same kernel, $V(H_0 - E)^{-1}\theta$, as in the IE for $T_1(E)$, cf. Eq. (16.38). Thus, by considering $V + V|0\rangle(E_0 - E)^{-1}\langle 0|T(E)$ as the new independent term, we can write that a solution for $T(E)$ must satisfy

$$T(E) = V + V|0\rangle(E_0 - E)^{-1}\langle 0|T(E) + K_1(E)\Big[V + V|0\rangle(E_0 - E)^{-1}\langle 0|T(E)\Big]$$

$$= T_1(E) + T_1(E)|0\rangle(E - E_0)^{-1}\langle 0|T(E) . \tag{16.42}$$

We now multiply the previous IE to the left by $\langle 0|$, so as to express $\langle 0|T(E)$ in terms of known matrix elements. It then results that

$$\langle 0|T(E) = \langle 0|T_1(E) + \langle 0|T_1(E)|0\rangle(E - E_0)^{-1}\langle 0|T(E) , \tag{16.43}$$

and then

$$\langle 0|T(E) = \Big[1 - \langle 0|T_1(E)|0\rangle(E_0 - E)^{-1}\Big]^{-1}\langle 0|T_1(E) . \tag{16.44}$$

Substituting the previous result into Eq. (16.42) we arrive to the final expression for $T(E)$,

$$T(E) = T_1(E) + T_1(E)|0\rangle\Big[E - E_0 - \langle 0|T_1(E)|0\rangle\Big]^{-1}\langle 0|T_1(E) . \tag{16.45}$$

From here we read the full propagator

$$\Delta(E) = \frac{1}{E - E_0 - \langle 0|T_1(E)|0\rangle} , \tag{16.46}$$

and the coupling squared operator $T_1(E)|0\rangle\langle 0|T_1(E)$. The latter when acting over the states in the continuum gives rise to the coupling function

$$\Theta(\mathbf{p}_n, E) = \langle \mathbf{p}_n|T_1(E)|0\rangle . \tag{16.47}$$

The Eq. (16.45) is the general expression of the scattering matrix $T(E)$ in terms of the reduced one $T_1(E)$, which results after a bare state $|0\rangle$ is removed from the sum over the intermediate states. In particular, it is clear that $T_1(E)$, cf. Eq. (16.38), corresponds to panel (a) of Fig. 16.1, $\Delta(E)$ in Eq. (16.46) arises from the Dyson resummation depicted in the panel (b), and the coupling function $\Theta(\mathbf{p}, E)$ of Eq. (16.47) originates from the FSI of the continuum states, drawn in the panel of (c) of the same figure.

The Eq. (16.31) is a particular case of Eq. (16.45) when projected over states in the continuum. The latter equation can be found in Ref. [100], but not its derivation, which has been offered here in detail for completeness and also for pedagogical reasons.

In particular, let us compare the expression in Eq. (16.5), given in terms of a subtraction constant and a CDD pole, with the model of Ref. [101] that results by applying Eq. (16.31) with $T_V(\mathbf{p}, \mathbf{p}', E)$ corresponding to the plain scattering length approximation,

$$T_V(\mathbf{p}, \mathbf{p}', E) = \frac{2\pi}{\mu} \frac{1}{-\frac{1}{a_V} - ik(E)} , \tag{16.48}$$

$$k(E) = \sqrt{2\mu E} .$$

For this particular case indeed $T_V(\mathbf{p}, \mathbf{p}', E)$ only depends on the energy and we better denote it simply as $T_V(E)$. As a result, the evaluation of the self-energy $G(E)$ and the dressed coupling $\Theta(\mathbf{p}, E)$, cf. Eqs. (16.34) and (16.35), respectively, is straightforward once $f(\mathbf{p})$ is known. Nonetheless, Ref. [101] argues that, since one is focusing in the low-energy region so that $k\alpha \ll 1$, with α the typical range of the interaction involved, one could parameterize the whole $f(\mathbf{p})$ by $f_0 = f(0)/(2\pi)$ and then, the diverging integrals from Eqs. (16.34) and (16.35) are regularized by naive dimensional analysis as

$$\tilde{g}_1(E) = \int \frac{d^3q}{(2\pi)^3} \frac{f(\mathbf{p})^2}{q^2/(2\mu) - E - i\varepsilon} = f_0^2(R + \mu ik) , \tag{16.49}$$

$$\tilde{g}_2(E) = \int \frac{d^3q}{(2\pi)^3} \frac{f(\mathbf{p})}{q^2/(2\mu) - E - i\varepsilon} = f_0(R' + \mu ik) ,$$

where one expects that the constants R and R' take values of $\mathcal{O}(\mu/\alpha)$. It is just a matter of simple algebra to deduce from Eqs. (16.34), (16.35) and (16.31) the following expression for the on-shell T matrix ($|\mathbf{p}| = |\mathbf{p}'| = k$) [101],

$$t(E) = -\frac{2\pi}{\mu} \frac{E - E_f + \frac{1}{2}g_f\gamma_V}{(E - E_f)(\gamma_V + ik) + i\frac{1}{2}g_f\gamma_V k} , \tag{16.50}$$

where $\gamma_V = 1/a_V$, while g_f and E_f are functions of the original parameters R, R' and E_0 of the model (the interested reader can consult Ref. [101] for the relations). Notice that g_f has the meaning of a bare coupling and E_f is the energy at which the real part of the denominator in Eq. (16.50) vanishes. Redefining the normalization multiplying $t(E)$ by $\mu/(2\pi)$, we end with a particular case of Eq. (16.5), previously obtained making use of general analytic and unitarity principles. The parameters in Eq. (16.5) are related to those in Eq. (16.50) by

$$\beta = -\gamma_V , \tag{16.51}$$

$$\gamma = \frac{1}{2} g_f \gamma_V^2 ,$$

$$M_{\text{CDD}} = E_f - \frac{1}{2} g_f \gamma_V .$$

However, the reverse is not true and Eq. (16.5) is not a particular case of Eq. (16.50) from the scattering model of Ref. [101]. As a proof of this statement, let us notice that from Eq. (16.50) the resulting effective range r can only be negative [98]. Applying Eq. (16.6) with the particular values of Eq. (16.51) we have

$$r = -\frac{g_f \gamma_V^2}{2\mu (E_f - g_f \gamma_V/2)^2} \leq 0 , \tag{16.52}$$

because $g_f = 2\mu f_0^2 (R - R_V)^2/R_V^2$ and $R_V = \mu \gamma_V$ [101].

The situation described in this section is particularly interesting for the scattering of heavy-quark mesons near their thresholds where several states with rather exotic properties have been found that go beyond well-established quarkonium spectroscopy [34, 102]. In these systems the coupling to the pion is relatively suppressed compared to that in the light-quark sector. For instance, for the P^*P potential worked out in Ref. [103] (where P^* is a heavy-quark vector-meson resonance and P is a heavy-quark pseudoscalar, with the heavy quark being the c or the b), one has that the strength of the central and tensor components of the one-pion mediated interaction is weaker by around a factor $g^2/(2g_A^2) \simeq 0.06$ compared to that for the NN interactions. Here g is the coupling for $P^*P\pi$ which is around 0.5 [103]. This makes that even though there would be a LHC due to pion exchanges, this can be treated perturbatively and one could neglect its effects in a first approximation. In such a situation we can then apply the results presented in this section [96, 98].

We have only introduced by pass the interesting matter of quantifying the compositeness and elementariness of a pole in the S matrix, since a full discussion on it should imply to abandon the strict realm of DRs and enter in QFT developments. For more discussions the interested reader can consult Ref. [14]. For earlier results one has, e.g., Refs. [96, 98–100, 104–108].

Chapter 17
An Example of Application of Analyticity in the Nuclear Medium: The Nuclear Energy Density

In this section, following Ref. [8], we apply DRs to calculate the nuclear-matter energy density, $\mathcal{E}$, as a function of the Fermi momenta for the protons and neutrons, ξ_p and ξ_n, respectively. Ref. [8] evaluates the contributions to the energy density of the nuclear medium up to and including NLO in the in-medium chiral counting developed in Ref. [109]. The different contributions are represented in Fig. 17.1. Without entering in the details of this in-medium chiral power counting, for what we refer to the latter reference, we focus our attention here to the contributions that generally stem from the iteration in the nuclear medium of the two-nucleon interactions, represented by the diagrams (c.1) and (c.2) in Fig. 17.1. The former corresponds to the direct NN interactions (Hartree diagrams) and the later to the crossed ones because of the Fermi statistics (Fock diagrams).

The contribution from the sum over the kinetic energies of the nucleons is given by the diagram (a) of Fig. 17.1, it is denoted by $\mathcal{E}_1$ and its expression is

$$\mathcal{E}_1 = \frac{3}{10m} \left(\rho_p \xi_p^2 + \rho_n \xi_n^2 \right) , \qquad (17.1)$$

where ρ_p and ρ_n are the proton and neutron densities. The latter read in terms of the corresponding Fermi momentum ξ_i,

$$\rho_i = 2 \int \frac{d^3 k}{(2\pi)^3} \theta(\xi_i - |\mathbf{k}|) = \frac{\xi_i^3}{3\pi^2} , \qquad (17.2)$$

with $i = 1(2)$ for the proton(neutron). The magnitude of $\mathcal{E}_1$ is suppressed with respect to its chiral order because it is divided by the relatively large nucleon mass. It is a contribution of recoil nature.

For the analysis of the rest of contributions in Fig. 17.1 we need to discuss the nucleon propagator in the nuclear medium with four-momentum k, $G_0(k)$. It can be written as [8, 110]

© The Author(s), under exclusive licence to Springer Nature Switzerland AG 2019
J. A. Oller, *A Brief Introduction to Dispersion Relations*, SpringerBriefs in Physics,
https://doi.org/10.1007/978-3-030-13582-9_17

$$G_0(k) = \left(\frac{1 + \tau_3}{2} \theta(\xi_p - |\mathbf{k}|) + \frac{1 - \tau_3}{2} \theta(\xi_n - |\mathbf{k}|) \right) \frac{1}{k^0 - E(\mathbf{k}) - i\epsilon}$$

$$+ \left(\frac{1 + \tau_3}{2} \theta(|\mathbf{k}| - \xi_p) + \frac{1 - \tau_3}{2} \theta(|\mathbf{k}| - \xi_n) \right) \frac{1}{k^0 - E(\mathbf{k}) + i\epsilon} . \quad (17.3)$$

In this expression $E(\mathbf{k})$ is the nucleon energy, $E(\mathbf{k}) = \sqrt{m^2 + \mathbf{k}^2}$ (we take the isospin limit for vacuum dynamics), and the τ_i are the Pauli matrices. We can also rewrite equivalently the nucleon propagator in Eq. (17.3) by doing the transformation $1/(x - i\varepsilon) = 1/(x + i\varepsilon) + 2i\pi\delta(x)$, with $x \to k^0 - E(\mathbf{k})$. It then reads,

$$G_0(k) = \frac{1}{k^0 - E(\mathbf{k}) + i\epsilon} \quad (17.4)$$

$$+ i(2\pi)\delta(k^0 - E(\mathbf{k})) \left(\frac{1 + \tau_3}{2} \theta(|\mathbf{k}| - \xi_p) + \frac{1 - \tau_3}{2} \theta(|\mathbf{k}| - \xi_n) \right) .$$

The first term is the free part of the propagator and the second one is the in-medium one. The latter one is also indicated as an in-medium insertion of a baryon propagator, or simply as an in-medium insertion. In Feynman diagrams an in-medium part of the nucleon propagator is depicted by a thick line, the free part by a line with a slash, and the full in-medium propagator is drawn by a plain line. The one-baryon propagator in Eqs. (17.3) and (17.4) is given in a matrix notation, while its components are denoted by $G_0(k)_i$.

The contribution (b) in Fig. 17.1, $\mathcal{E}_2$, arises from the nucleon self-energy due to a pion loop. It entails only one in-medium insertion, because a contribution with two in-medium insertions is already accounted for by the diagram (c.2) [due to the isovector nature of the pion–nucleon coupling there is no one-pion loop (c.1)-like diagram for this case]. We denote by Σ_f^π the nucleon self-energy in vacuum by a pion loop, which expression reads [8, 111]

$$\Sigma_f^\pi(k) = \frac{3g_A^2 b}{32\pi^2 f_\pi^2} \left[-\omega + \sqrt{b} \left(i \log \frac{\omega + i\sqrt{b}}{-\omega + i\sqrt{b}} + \pi \right) \right] - \frac{3g_A^2 m_\pi^3}{32\pi f_\pi^2} , \quad (17.5)$$

with $g_A \simeq 1.26$ the axial coupling of the nucleon related by chiral symmetry (partially conserved axial-vector current) to the pion–nucleon coupling constant. In the previous equation $\omega = k^0$ is the nucleon energy once its rest mass is discounted and $b = m_\pi^2 - \omega^2 - i\varepsilon$. The last term in Eq. (17.5) is subtracted because the self-energy is zero for the $\omega = 0$, which corresponds to the vacuum nucleon mass at rest. Since in the diagram (b) of Fig. 17.1 the nucleon energy is a kinetic one, with $\xi_i \ll m$, it follows that this diagram is indeed a small contribution to the total energy density in the medium.

For evaluating the contributions (c.1) and (c.2) of Fig. 17.1 we need the in-medium NN interactions, that are depicted by the iteration of the zig-zag lines. For an in-medium NN PWA we use Eq. (7.2) in terms of $\mathcal{N}$, that only has LHC, and the two-nucleon unitary function, that in the nuclear medium corresponds to $L_{10}^{l_3}$ instead

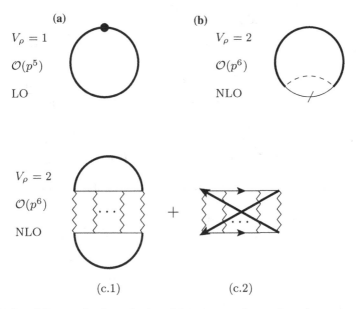

Fig. 17.1 Set of diagrams for the evaluation of the energy per baryon in nuclear matter up to an including two-nucleon interactions in the nuclear medium. In-medium insertions are represented in the figure by thick solid lines, and the thin ones correspond to the full baryon propagator $G_0(k)$, cf. Eq. (17.4). The diagram (**a**) is the kinetic energy, (**b**) represents the nucleon self-energy due to a pion loop [it involves one in-medium and one free baryon propagator (solid line with a dash), so as not to double count with the diagrams in (**c**)]. Finally, diagrams (c.1) (Hartree) and (c.2) (Fock diagrams) are the contributions due to the direct and exchange two-nucleon interactions, in order, with at least two in-medium interactions in the baryon propagators. Its evaluation [8], by making use of a partial-wave expansion and the analytical properties of the PWAs in the nuclear medium, is the main point of the present section

of $g(s)$. Contrarily to the vacuum case, the in-medium unitarity loop function function also depends on the total CM three-momentum of the two nucleons. In terms of the four-momenta k_1 and k_2 of the two nucleons we introduce the four-momenta a and p defined as

$$a = \frac{1}{2}(k_1 + k_2) , \tag{17.6}$$

$$p = \frac{1}{2}(k_1 - k_2) .$$

We also use below the quantity

$$A = 2ma^0 - \mathbf{a}^2 . \tag{17.7}$$

The two-nucleon unitarity function depends also on the total charge of the two nucleons because the different values that the Fermi momenta of protons and nucleons could have. This is indicated by the superscript I_3 in $L_{10}^{I_3}$ which corresponds to the total third component of the isospin of the NN system. The explicit expression for $L_{10}^{I_3}$ is [8]

$$L_{10}^{I_3} = i \int \frac{d^4k}{(2\pi)^4} \left[\frac{\theta(\xi_1 - |\mathbf{a} - \mathbf{k}|)}{a^0 - k^0 - E(\mathbf{a} - \mathbf{k}) - i\epsilon} + \frac{\theta(|\mathbf{a} - \mathbf{k}| - \xi_1)}{a^0 - k^0 - E(\mathbf{a} - \mathbf{k}) + i\epsilon} \right]$$
$$\times \left[\frac{\theta(\xi_2 - |\mathbf{a} + \mathbf{k}|)}{a^0 + k^0 - E(\mathbf{a} + \mathbf{k}) - i\epsilon} + \frac{\theta(|\mathbf{a} + \mathbf{k}| - \xi_2)}{a^0 + k^0 - E(\mathbf{a} + \mathbf{k}) + i\epsilon} \right] . \tag{17.8}$$

Performing explicitly the integration over k^0 we have for this loop function

$$L_{10}^{I_3} = m \int \frac{d^3k}{(2\pi)^3} \left[\frac{\theta(|\mathbf{a} - \mathbf{k}| - \xi_1)\theta(|\mathbf{a} + \mathbf{k}| - \xi_2)}{A - \mathbf{k}^2 + i\epsilon} \right.$$
$$\left. - \frac{\theta(\xi_1 - |\mathbf{a} - \mathbf{k}|)\theta(\xi_2 - |\mathbf{a} + \mathbf{k}|)}{A - \mathbf{k}^2 - i\epsilon} \right], \tag{17.9}$$

in which the first term between square brackets is the free particle-free particle part (in the following we drop the adjective "free" as usual in the literature) and the last one is the so-called hole–hole part (because it involves two insertions of Fermi seas due to the Heaviside functions in the numerator). The integration over $\mathbf{k}$ can also be performed algebraically and the explicit expressions can be found in the Appendix C of Ref. [8]. It is clear from Eq. (17.9) the dependence of $L_{10}^{I_3}$ on I_3 and the CM variables contained in a and A. In this respect, notice that in the CM frame and for on-shell k_1 and k_2, it follows from Eq. (17.7) that $A = \mathbf{p}^2$. The Eq. (17.9) also establishes the appearance of the RHC when the real part of any of its denominators vanishes. The resulting imaginary part has the same sign from both contributions because of the minus sign in front of the hole–hole term. At LO in the in-medium chiral counting of Ref. [109] the matrix $\mathcal{N}$ is the same as the one already determined in vacuum. In general its characteristic facet at any order in the chiral expansion, as expressed above, is that it has no RHC, being the latter contained entirely in $L_{10}^{I_3}$. Employing the notation of Ref. [8] we denote the former by $\mathcal{N}_{JI}(\bar{\ell}, \ell, S)$ and similarly for the PWA, $T_{JI}^{I_3}(\bar{\ell}, \ell, S)$. Here, J is the total angular momentum, S the total spin, $\bar{\ell}$ the final orbital angular momentum and ℓ the initial one, always referred to the initial/final NN systems (the meaning of the different labels is in harmony with the notation introduced in Chap. 2). After this preamble, the in-medium expression equivalent to Eq. (7.2) is

$$T_{JI}^{I_3}(\bar{\ell}, \ell, S; \mathbf{p}^2, \mathbf{a}^2, A) = \left[\mathcal{N}_{JI}^{I_3}(\bar{\ell}, \ell, S; \mathbf{p}^2, \mathbf{a}^2, A)^{-1} + L_{10}^{I_3}(\mathbf{a}^2, A) \right]^{-1} . \tag{17.10}$$

It is worth clarifying that the total isospin I of a NN state is a good quantum number because the $L_{10}^{I_3}$ function is symmetric under the exchange of the two particles, cf. Eq. (17.8). This is a general rule because the $I_3 = 0$ operators are symmetric under the exchange $p \leftrightarrow n$, and therefore the symmetric properties under the transposition of the two particles in the $I_3 = 0 \, NN$ state are not altered by the iterative interacting process.

Let us come back to evaluate the diagrams (c.1) and (c.2) in Fig. 17.1. Its sum is denoted by $\mathcal{E}_3$ and it is given by

$$
\mathcal{E}_3 = \frac{1}{2} \sum_{\sigma_1, \sigma_2} \sum_{\alpha_1, \alpha_2} \int \frac{d^4 k_1}{(2\pi)^4} \frac{d^4 k_2}{(2\pi)^4} e^{i k_1^0 \eta} e^{i k_2^0 \eta} G_0(k_1)_{\alpha_1} G_0(k_2)_{\alpha_2} \tag{17.11}
$$
$$
\times \, T_{\alpha_1 \alpha_2}^{\sigma_1 \sigma_2} (\mathbf{p}, \mathbf{a}, A) \,.
$$

In this equation $\eta \to 0^+$ at the end of the calculation. It is introduced so as to enforce that at least two in-medium insertions get involved in the calculation [8, 110]. The NN scattering amplitude from the initial state $|k_1, k_2, \sigma_1 \sigma_2, \alpha_1 \alpha_2\rangle_S$, cf. Eq. (2.55), to the same final one is indicated in the previous equation by $T_{\alpha_1 \alpha_2}^{\sigma_1 \sigma_2} (\mathbf{p}, \mathbf{a}, A)$. The two states are the same because one has to take the trace of the scattering amplitudes when calculating the self-interactions of the system giving rise to $\mathcal{E}_3$. As in Chap. 2, the labels σ_i and α_i refer to the third components of the spin and isospin of the ith nucleon, in order. Since Eq. (17.11) is already a NLO contribution we can use for its evaluation the LO NN PWAs amplitudes, by employing Eq. (17.10) with $\mathcal{N}$ calculated as in vacuum. In such a case, $\mathcal{N}_{JI}(\bar{\ell}, \ell, S)$ is a function only of $\mathbf{p}^2$, the momentum transfer squared, cf. Eqs. (17.6), and (17.10) becomes

$$
T_{JI}^{I_3}(\bar{\ell}, \ell, S; \mathbf{p}^2, \mathbf{a}^2, A) = \left[\mathcal{N}_{JI}^{I_3}(\bar{\ell}, \ell, S; \mathbf{p}^2)^{-1} + L_{10}^{I_3}(\mathbf{a}^2, A) \right]^{-1} . \tag{17.12}
$$

Once the integration variables k_1 and k_2 are changed by A, $\mathbf{a}$ and p in Eq. (17.11), it is then possible to perform straightforwardly the integration over p^0. Notice that the only dependence on p^0 in the integrand of Eq. (17.11) is in the propagators $G_0(k_i)_{\alpha_i}$. It results,

$$
\int \frac{dp^0}{2\pi} G_0(a+p)_{\alpha_1} G_0(a-p)_{\alpha_2} =
$$
$$
- i \left[\frac{\theta(|\mathbf{a}+\mathbf{p}| - \xi_{\alpha_1}) \theta(|\mathbf{a}-\mathbf{p}| - \xi_{\alpha_2})}{2a^0 - E(\mathbf{a}+\mathbf{p}) - E(\mathbf{a}-\mathbf{p}) + i\epsilon} - \frac{\theta(\xi_{\alpha_1} - |\mathbf{a}+\mathbf{p}|) \theta(\xi_{\alpha_2} - |\mathbf{a}-\mathbf{p}|)}{2a^0 - E(\mathbf{a}+\mathbf{p}) - E(\mathbf{a}-\mathbf{p}) - i\epsilon} \right] .
$$
$$
\tag{17.13}
$$

For convenience we also introduce the splitting of the particle–particle contribution in the form

$$
\theta(|\mathbf{a}+\mathbf{p}| - \xi_{\alpha_1}) \theta(|\mathbf{a}-\mathbf{p}| - \xi_{\alpha_2}) = [1 - \theta(\xi_{\alpha_1} - |\mathbf{a}+\mathbf{p}|)][1 - \theta(\xi_{\alpha_2} - |\mathbf{a}-\mathbf{p}|)]
$$
$$
= 1 - \theta(\xi_{\alpha_1} - |\mathbf{a}+\mathbf{p}|) - \theta(\xi_{\alpha_2} - |\mathbf{a}-\mathbf{p}|) + \theta(\xi_{\alpha_1} - |\mathbf{a}+\mathbf{p}|) \theta(\xi_{\alpha_2} - |\mathbf{a}-\mathbf{p}|) .
$$
$$
\tag{17.14}
$$

It follows from Eqs. (17.13) and (17.14) that the result of the integration in p^0 of Eq. (17.11) can be written as

$$
\mathcal{E}_3 = -4i \sum_{\sigma_1,\sigma_2} \sum_{\alpha_1,\alpha_2} \int \frac{d^3a}{(2\pi)^3} \frac{d^3p}{(2\pi)^3} \frac{dA}{2\pi} e^{i(A+a^2)\eta} T^{\sigma_1\sigma_2}_{\alpha_1\alpha_2}(\mathbf{p},\mathbf{a},A) \left[\frac{1}{A - \mathbf{p}^2 + i\epsilon} \right.
$$

$$
- \frac{\theta(\xi_{\alpha_1} - |\mathbf{a}+\mathbf{p}|) + \theta(\xi_{\alpha_2} - |\mathbf{a}-\mathbf{p}|)}{A - \mathbf{p}^2 + i\epsilon}
$$

$$
\left. - 2\pi i \delta(A - \mathbf{p}^2)\theta(\xi_{\alpha_1} - |\mathbf{a}+\mathbf{p}|)\theta(\xi_{\alpha_2} - |\mathbf{a}-\mathbf{p}|) \right]. \tag{17.15}
$$

Here we have expressed all the explicit denominators having $+i\epsilon$ by employing the trick explained just before Eq. (17.4).

The next step is to perform the integration over A, which actually implies to compute

$$
\int_{-\infty}^{\infty} \frac{dA}{2\pi} \frac{e^{iA\eta}}{A - \mathbf{p}^2 + i\epsilon} T^{\sigma_1\sigma_2}_{\alpha_1\alpha_2}(\mathbf{p},\mathbf{a},A) , \tag{17.16}
$$

since the last term in the integrand of Eq. (17.15) is proportional to $\delta(A - \mathbf{p}^2)$ and the integral in A is then trivial. We proceed with Eq. (17.16) by enclosing the integration contour in A with a semicircle at infinity in the half complex A plane with positive imaginary part, by taking advantage of the factor $e^{iA\eta}$ with $\eta \to 0^+$. As it is evident from Eq. (17.9), the particle–particle contribution gives rise to a cut in A with a slightly negative imaginary part, so that it is not within the domain that results after closing the integration contour. Similarly the denominator in Eq. (17.16) gives rise to a pole singularity in A with also a negative imaginary part. However, the hole–hole part in $L^{I_3}_{10}(\mathbf{a},A)$ generates a cut in A than runs slightly above the real axis with a positive imaginary part. This cut is of finite extent because of the Heaviside functions in the hole–hole part and extends from $A_1(|\mathbf{a}|)$ up to $A_2(|\mathbf{a}|)$ as depicted in Fig. 17.2 by the dashed line (explicit expressions for these limits are given in Eq. (C.19) of Ref. [8].)[1]

In order to go on and perform the integration in A we proceeds as follows. We consider two closed contours in the form stated above, but one of them runs above the hole–hole cut and the other below it. The former integration contour is denoted by $\mathcal{C}_{I'}$, the latter by $\mathcal{C}_I$, and both are represented in Fig. 17.2. We have the following preliminary results,

---

[1] These limits depend on I_3, although this is not explicitly written, since no ambiguity arises once the partial-wave expansion of the T matrix is performed below.

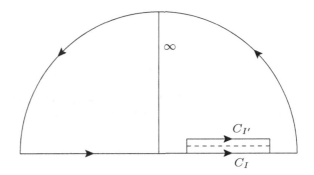

Fig. 17.2 Integration contours used to evaluate the A integration in Eq. (17.16). The dashed line is the RHC due to the hole–hole contribution in $L_{10}^{l_3}(\mathbf{a}^2, A)$

$$\int_{-\infty}^{\infty} \frac{dA}{2\pi} \frac{e^{iA\eta}}{A - \mathbf{p}^2 + i\epsilon} T_{\alpha_1\alpha_2}^{\sigma_1\sigma_2}(\mathbf{p}, \mathbf{a}, A) = \oint_{C_I} \frac{dA}{2\pi} \frac{e^{iA\eta}}{A - \mathbf{p}^2 + i\epsilon} T_{\alpha_1\alpha_2}^{\sigma_1\sigma_2}(\mathbf{p}, \mathbf{a}, A) ,$$

$$\oint_{C_{I'}} \frac{dA}{2\pi} \frac{e^{iA\eta}}{A - \mathbf{p}^2 + i\epsilon} T_{\alpha_1\alpha_2}^{\sigma_1\sigma_2}(\mathbf{p}, \mathbf{a}, A) = 0 . \tag{17.17}$$

Therefore, the subtraction of the two integrals gives

$$\oint_{C_I} \frac{dA}{2\pi} \frac{e^{iA\eta}}{A - \mathbf{p}^2 + i\epsilon} - \oint_{C_{I'}} \frac{dA}{2\pi} \frac{e^{iA\eta}}{A - \mathbf{p}^2 + i\epsilon}$$
$$= \int_{A_1(|\mathbf{a}|)}^{A_2(|\mathbf{a}|)} \frac{dA}{2\pi} \frac{T_{\alpha_1\alpha_2}^{\sigma_1\sigma_2}(\mathbf{p}, \mathbf{a}, A) - T_{\alpha_1\alpha_2}^{\sigma_1\sigma_2}(\mathbf{p}, \mathbf{a}, A + 2i\epsilon)}{A - \mathbf{p}^2 + i\epsilon} . \tag{17.18}$$

Notice that this result is also a consequence of deforming the integration contour C_I for avoiding the cut.

An interesting result in Ref. [8] is the derivation of the partial-wave expansion of the NN scattering amplitude in the nuclear medium, despite its dependence on $\mathbf{a}$. This is a generalization of the results in Chap. 2. The scattering amplitude $T_{\alpha_1\alpha_2}^{\sigma_1\sigma_2}(\mathbf{p}, \mathbf{a}, A)$ in terms of the in-medium NN PWAs, by making use of Eq. (A.8) of Ref. [8], reads

$$T_{\alpha_1\alpha_2}^{\sigma_1\sigma_2} = 4\pi \sum (\sigma_1\sigma_2 s_3|s_1 s_2 S)^2 (m's_3\mu|\ell'SJ)(ms_3\mu|\ell SJ)(\alpha_1\alpha_2 i_3|\tau_1\tau_2 I)^2 \tag{17.19}$$
$$\times Y_{\ell'}^{m'}(\hat{\mathbf{p}}')Y_\ell^m(\hat{\mathbf{p}})^* \chi(S\ell'I)\chi(S\ell I)T_{JI}^{l_3}(\ell', \ell, S) .$$

For NN scattering $\tau_1 = \tau_2 = s_1 = s_2 = 1/2$ and the symbol $\chi(S\ell I)$ arises because Fermi statistics and it is

$$\chi(S\ell I) = \frac{1 - (-1)^{\ell+S+I}}{\sqrt{2}} = \begin{cases} \sqrt{2} & \ell + S + I = \text{odd} , \\ 0 & \ell + S + I = \text{even} . \end{cases} \tag{17.20}$$

This factor accounts for the unitary normalization introduced in Chap. 2.

The sum over the isospin and spin indices is straightforward,

$$\sum_{\alpha_1\alpha_2}(\alpha_1\alpha_2 i_3|\tau_1\tau_2 I)^2 = 1 \, , \tag{17.21}$$

$$\sum_{\sigma_1\sigma_2}(\sigma_1\sigma_2 s_3|s_1 s_2 S)^2 = 1 \, .$$

We continue next with the sum over s_3 and the third components of orbital angular momentum, which can also be summed in a close form as

$$\sum_{m',m,s_3} (m's_3\mu|\ell'SJ)(ms_3\mu|\ell SJ)Y_{\ell'}^{m'}(\hat{\mathbf{p}})Y_\ell^m(\hat{\mathbf{p}})^* = \delta_{\ell'\ell}\frac{2J+1}{4\pi} \, . \tag{17.22}$$

To arrive to this result we have used the following symmetry property of the Clebsch–Gordan coefficients [6],

$$(m_1 m_2 m_3|j_1 j_2 j_3) = (-1)^{m_2+j_2}\left(\frac{2j_3+1}{2j_1+1}\right)^{1/2}(-m_2 m_3 m_1|j_2 j_3 j_1) \, . \tag{17.23}$$

This property allows us to write the sum of Clebsch–Gordan coefficients in Eq. (17.22) as

$$\sum_{s_3,\mu}(m's_3\mu|\ell'SJ)(ms_3\mu|\ell SJ) = \frac{2J+1}{\sqrt{(2\ell+1)(2\ell'+1)}}\sum_{s_3,\mu}(-s_3\mu m'|S\ell'J)$$

$$\times (-s_3\mu m|S\ell J) = \frac{2J+1}{2\ell+1}\delta_{\ell'\ell}\delta_{m'm'} \, . \tag{17.24}$$

Notice that we could have included a sum over μ already in Eq. (17.22), because μ is fixed by the properties of the Clebsch–Gordan coefficients. Finally, the result in Eq. (17.22) follows by employing the addition theorem of the spherical harmonics

$$\frac{1}{2\ell+1}\sum_m |Y_\ell^m(\hat{\mathbf{p}})|^2 = \frac{1}{4\pi} \, . \tag{17.25}$$

Thus, the sum over the PWAs for calculating $\mathcal{E}_3$ simplifies to

$$\sum_{\alpha_1,\alpha_2}\sum_{\sigma_1,\sigma_2} T_{\alpha_1\alpha_2}^{\sigma_1\sigma_2}(\mathbf{p},\mathbf{a},A) = \sum_{I,I_3,J,\ell,S}(2J+1)\chi(S\ell I)^2 T_{JI}^{I_3}(\ell,\ell,S;\mathbf{p}^2,\mathbf{a}^2,A) \, . \tag{17.26}$$

In the rest of this section we suppress the arguments (ℓ,ℓ,S) in $T_{JI}^{I_3}$ and $\mathcal{N}_{JI}^{I_3}$ for brevity in the writing. We now perform the difference between PWAs needed to implement Eq. (17.18),

$$T_{JI}^{I_3}(\mathbf{p}^2, \mathbf{a}^2, A) - T_{JI}^{I_3}(\mathbf{p}^2, \mathbf{a}^2, A + i2\epsilon) = \left[N_{JI}^{I_3}(\mathbf{p}^2) + L_{10}^{I_3}(\mathbf{a}^2, A)\right]^{-1}$$

$$- \left[N_{JI}^{I_3}(\mathbf{p}^2) + L_{10}^{I_3}(\mathbf{a}^2, A + i2\epsilon)\right]^{-1} = \left[N_{JI}^{I_3}(\mathbf{p}^2) + L_{10}^{I_3}(\mathbf{a}^2, A)\right]^{-1}$$

$$\times \left[L_{10}^{I_3}(\mathbf{a}^2, A + i2\epsilon) - L_{10}^{I_3}(\mathbf{a}^2, A)\right]\left[N_{JI}^{I_3}(\mathbf{p}^2) + L_{10}^{I_3}(\mathbf{a}^2, A + i2\epsilon)\right]^{-1}. \quad (17.27)$$

It follows from Eq. (17.9) that the difference $L_{10}^{I_3}(\mathbf{a}^2, A + i2\epsilon) - L_{10}^{I_3}(\mathbf{a}^2, A)$ is due entirely to the hole–hole part and it gives

$$L_{10}^{I_3}(\mathbf{a}^2, A + i2\epsilon) - L_{10}^{I_3}(\mathbf{a}^2, A) = -m \int \frac{d^3q}{(2\pi)^3} \theta(\xi_{\alpha_1} - |\mathbf{a} + \mathbf{q}|)\theta(\xi_{\alpha_2} - |\mathbf{a} - \mathbf{q}|)$$

$$\times \left(\frac{1}{A - \mathbf{q}^2 + i\epsilon} - \frac{1}{A - \mathbf{q}^2 - i\epsilon}\right)$$

$$= i2\pi m \int \frac{d^3q}{(2\pi)^3} \theta(\xi_{\alpha_1} - |\mathbf{a} + \mathbf{q}|)\theta(\xi_{\alpha_2} - |\mathbf{a} - \mathbf{q}|)\delta(A - \mathbf{q}^2). \quad (17.28)$$

Now, we come back to Eq. (17.15) and from Eqs. (17.26), (17.27) and (17.28) it follows that after performing the integration in A, cf. Eq. (17.18), we can write Eq. (17.15) as

$$\mathcal{E}_3 = -4 \sum_{I, I_3, J, \ell, S} (2J + 1)\chi(S\ell I)^2 \int \frac{d^3a}{(2\pi)^3} \frac{d^3q}{(2\pi)^3} \theta(\xi_{\alpha_1} - |\mathbf{a} + \mathbf{q}|)\theta(\xi_{\alpha_2} - |\mathbf{a} - \mathbf{q}|)$$

$$\times \left(T_{JI}^{I_3}(\mathbf{q}^2, \mathbf{a}^2, \mathbf{q}^2) + m \int \frac{d^3p}{(2\pi)^3} \frac{1 - \theta(\xi_{\alpha_1} - |\mathbf{a} + \mathbf{p}|) - \theta(\xi_{\alpha_2} - |\mathbf{a} - \mathbf{p}|)}{\mathbf{p}^2 - \mathbf{q}^2 - i\epsilon}\right.$$

$$\times \left[\mathcal{N}_{JI}^{I_3}(\mathbf{p}^2)^{-1} + L_{10}^{I_3}(\mathbf{a}^2, \mathbf{q}^2)\right]^{-1}\left[\mathcal{N}_{JI}^{I_3}(\mathbf{p}^2)^{-1} + L_{10}^{I_3}(\mathbf{a}^2, \mathbf{q}^2)^*\right]^{-1}\Bigg)_{(\ell, \ell, S)}. \quad (17.29)$$

This is our final equation for $\mathcal{E}_3$. In this equation we have dropped the exponent $e^{i\mathbf{a}^2\eta}$ since the integration in $|\mathbf{a}|$ is bounded because of the product of the Heaviside functions $\theta(\xi_{\alpha_1} - |\mathbf{a} + \mathbf{q}|)\theta(\xi_{\alpha_2} - |\mathbf{a} - \mathbf{q}|)$. Related to this factor, we have also written that $L_{10}^{I_3}(\mathbf{a}^2, \mathbf{q}^2 + i2\epsilon) = L_{10}^{I_3}(\mathbf{a}^2, \mathbf{q}^2)^*$, because only the hole–hole part in this functions enters, cf. Eq. (17.9). The rest of sums and integrations in Eq. (17.29) are performed numerically in Ref. [8].

It is also of pedagogical interest to show explicitly following Ref. [8] that $\Im\mathcal{E}_3 = 0$, as it must be because $\mathcal{E}$ is a real quantity. The imaginary part of the second term between the round brackets in Eq. (17.29) stems only from the denominator of $1/(\mathbf{p}^2 - \mathbf{q}^2 - i\epsilon) \to i\pi\delta(\mathbf{p}^2 - \mathbf{q}^2)$. We then have

$$\Im\mathcal{E}_3 = -4 \sum_{I,I_3,J,\ell,S} (2J+1)\chi(S\ell I)^2 \int \frac{d^3a}{(2\pi)^3} \frac{d^3q}{(2\pi)^3} \theta(\xi_{\alpha_1} - |\mathbf{a}+\mathbf{q}|)\theta(\xi_{\alpha_2} - |\mathbf{a}-\mathbf{q}|)$$

$$\times \left(\Im T_{JI}^{I_3}(q^2, a^2, q^2) + m \int \frac{d^3p}{(2\pi)^3}[1 - \theta_{\alpha_1}(\xi_{\alpha_1} - |\mathbf{a}+\mathbf{p}|) - \theta_{\alpha_2}(\xi_{\alpha_2} - |\mathbf{a}-\mathbf{p}|)] \right.$$

$$\left. \times \pi\delta(p^2 - q^2)T_{JI}^{I_3}(p^2, a^2, q^2)T_{JI}^{I_3}(p^2, a^2, q^2)^* \right)_{(\ell,\ell,S)}. \tag{17.30}$$

It is clear from Eq. (17.12) (with $A = q^2$) that the imaginary part of $T_{JI}^{I_3}(q^2, a^2, q^2)$ arises from the one of $L_{10}^{I_3}(a^2, q^2)$, which in turn is only due to the hole–hole part because of the product of the two Heaviside functions on the rhs of Eq. (17.30). Substituted the expression for $\Im T_{JI}^{I_3}$ into the previous equation one finds

$$\Im\mathcal{E}_3 = -4 \sum_{I,I_3,\ell,S} (2J+1)\chi(S\ell I)^2 \int \frac{d^3a}{(2\pi)^3} \frac{d^3q}{(2\pi)^3} \theta(\xi_{\alpha_1} - |\mathbf{a}+\mathbf{q}|)\theta(\xi_{\alpha_2} - |\mathbf{a}-\mathbf{q}|)$$

$$\times \int \frac{d^3p}{(2\pi)^3} m\pi\delta(p^2 - q^2)\big[1 - \theta(\xi_{\alpha_1} - |\mathbf{a}+\mathbf{p}|) - \theta(\xi_{\alpha_2} - |\mathbf{a}-\mathbf{p}|)$$

$$+ \theta(\xi_{\alpha_1} - |\mathbf{a}+\mathbf{p}|)\theta(\xi_{\alpha_2} - |\mathbf{a}-\mathbf{p}|)\big]T_{JI}^{I_3}T_{JI}^{I_3*}\Big|_{(\ell,\ell,S)} = 0. \tag{17.31}$$

To conclude that this expression is zero, we have taken into account from Eq. (17.14) that the function between square brackets in the previous equation is only the particle–particle part, given by $\theta(|\mathbf{a}+\mathbf{p}| - \xi_{\alpha_1})\theta(|\mathbf{a}-\mathbf{p}| - \xi_{\alpha_2})$. But since $|\mathbf{p}| = |\mathbf{q}|$ there is no way that this product of step functions can be satisfied because $\mathbf{a}$ and $\mathbf{q}$ are already constrained to satisfy the product of the two Heaviside functions in the first line of Eq. (17.31). Thus, this equation is zero.

Appendix A
Numerical Method to Solve
the Coupled Linear IEs in Eq. (15.23)

We now discuss briefly how Ref. [93] proceeds to solve numerically the IEs of Eq. (15.23). It also introduces a general interesting method to deal with the Cauchy's principal value prescription, which is very often found in actual applications of DRs. This method was exposed originally in the Appendix C of the eprint version [112].

To explain the method it is enough to consider the reduction of Eq. (15.23) to the uncoupled case. After using the well-known result that under an integration over x one has that $1/(x \pm i\varepsilon) = $ Principal Value $\pm i\pi\delta(x)$, we write

$$\Re F(s) = \frac{1}{\pi} \fint_{s_{\mathrm{th}}}^{\infty} \frac{f(s')ds'}{s' - s} , \tag{A.1}$$

$$f(s) = \Im F(s) = \rho(s)F(s)T(s)^* .$$

Here $s_{\mathrm{th}} = m_K^2 + m_\pi^2$ and the integral symbol with the dash refers to the Cauchy's principal value. The integration is split in two pieces by introducing $s_{\mathrm{cut}} > s_{\mathrm{th}}$, such that the first integral comprises the energy interval in which the form factor has more structure (and therefore it requires a higher numerical load). Next, we perform a change of variables to x so that the range of integration in both integrals is $[0, 1]$. For the first integral $s' \in [s_{\mathrm{th}}, s_{\mathrm{cut}}]$ the variable x is defined as

$$x = \sqrt{\frac{s' - s_{\mathrm{th}}}{s_{\mathrm{cut}} - s_{\mathrm{th}}}} . \tag{A.2}$$

For the second integral $s' \in [s_{\mathrm{cut}}, \infty)$ the new integration variable z is

$$z = \frac{s' - s_{\mathrm{th}}}{s' - b\, s_{\mathrm{th}}} , \tag{A.3}$$

with $b < 1$. This parameter is introduced to test the numerical stability of the method as well as for increasing the numerical convergence for reaching the result. In this way, Eq. (A.1) is recast as

$$\Re F(s) = \int_0^1 \frac{2x' f(s'(x'))dx'}{x'^2 - x^2} + \int_0^1 \frac{(1-z')f(s'(z'))dz'}{(1-z')(z'-z)} . \tag{A.4}$$

Now, let us discuss the numerical algorithm to perform the Cauchy's principal value integrals. We employ the Gaussian quadrature algorithm to integrate the prototype integral

$$I(x) = \int_{-1}^1 \frac{f(y)dy}{x - y} , \tag{A.5}$$

where the weight function $w(y) = 1/(x - y)$ is the appropriate one for the integral in Eq. (A.1). First we have to find a set of orthogonal functions $u_n(y; x)$ given the weight function, such that

$$\langle u_n | u_m \rangle = \int_{-1}^1 \frac{u_n(y; x)u_m(y; x)dy}{x - y} = 0 . \tag{A.6}$$

One can easily check that a set of orthogonal function is given by

$$u_n(y; x) = P_n(y) - \frac{Q_n(x)}{Q_{n-1}(x)} P_{n-1}(y) , \tag{A.7}$$

where the $P_n(y)$ are the Legendre polynomials and the $Q_n(x)$ are the associated Legendre function of the second kind (Chap. 7.22 of Ref. [113]). The normalization can be also easily found with the same techniques,

$$\langle u_n | u_n \rangle = 2Q_n(x)u_n(x; x) . \tag{A.8}$$

For implementing a Gaussian algorithm of order N for evaluating $I(x)$, the set of points in which $f(y)$ is evaluated to perform the integrals corresponds to the zeroes of $u_N(y; x)$ (with respect to the argument y). This problem in principle gets complicated because of the fact that the orthogonal function $u_N(y; x)$ depends also on x. To avoid that, one first evaluate the integral along the points x_i that are zero of $Q_N(x)$, $Q_N(x_i) = 0$. For such values the orthogonal function is just the Legendre polynomial. The weights in the numerical integral

$$I(x_i) = \sum_{j=1}^N w_j(x_i) f(y_j) , \tag{A.9}$$

can be evaluated from a generic formula given in the Numerical Recipes [114], with the result

$$w_j(x_i) = \frac{2(1 - y_j^2)}{N(x_i - y_i)} \frac{Q_{N-1}(x_i)P_N(x_i)}{P_{N-1}(y_j)^2} . \tag{A.10}$$

The integral $I(x)$ for other values of x $(x \neq x_i)$ can be obtained by employing interpolating algorithms [114]. It is also worth indicating that the zeros of $P_N(x)$, $x \in [-1, 1]$, can also be found making use of standard algorithms [114]. The finding of the $N + 1$ roots of $Q_N(x)$ get simplified if one takes into account that they interleave the roots of $P_N(x)$.

Let us indicate that in order to calculate the integral for a generic interval $z \in (a, b)$, with the integration variable z, one can make the linear transformation $y = (2z - a - b)/(b - a)$, so that $y \in [-1, 1]$. Finally, in the Appendix D of Ref. [112] one can find a Fortran implementation of this method which is based on the Gauss–Legendre algorithm that can be found in the Numerical Recipes [114].

References

1. Heisenberg, W.: Z. Phys. **120**, 513 (1943)
2. Eden, R.J., Landshoff, P.V., Olive, D.I., Polkinghorne, J.C.: The Analytic S-Matrix. Cambridge University Press, Cambridge (1966)
3. Henley, E.M., Thirring, W.: Elementary Quantum Field Theory, McGraw-Hill Book Company, Inc., New York, (1962) (Chapter 5.4)
4. Weinberg, S.: The Quantum Field Theory of Fields. Foundations, vol. I. Cambridge University Press, New York (1995)
5. Martin, A.D., Spearman, T.D.: Elementary Particle Theory. North-Holland Publishing Company, Amsterdam (1970)
6. Rose, M.E.: Elementary Theory of Angular Momentum. Dover, New York (1995)
7. Oller, J.A., Entem, D.R. arXiv:1810.12242 [hep-ph]
8. Lacour, A., Oller, J.A., Meißner, U.-G.: Ann. Phys. **326**, 241 (2011)
9. Gülmez, D., Meißner, U.-G., Oller, J.A.: Eur. Phys. J. C **77**, 460 (2017)
10. Oller, J.A., Oset, E.: Nucl. Phys. A **620**, 438 (1997); (E) *ibid.* **652**, 407, (1999)
11. Meißner, U.G., Oller, J.A.: Nucl. Phys. A **679**, 671 (2001)
12. Alarcon, J.M., Martin Camalich, J., Oller, J.A.: Ann. Phys. **336**, 413 (2013)
13. Chew, G.F., Goldberger, M.L., Low, F.E., Nambu, Y.: Phys. Rev. **106**, 1337 (1957)
14. Oller, J.A.: Ann. Phys. **396**, 429 (2018)
15. Sugawara, M., Kanazawa, A.: Phys. Rev. **123**, 1895 (1961)
16. Herglotz, A.: Ver. Verhandl. Sachs. Ges. Wiss. Leipzig, Math. Phys. **63**, 501 (1911); Shohat, J.A., Tamarkin, J.D.: The Problem of Moments. American Mathematical Society, p. 24. New York, (1943), chapter II
17. Weinberg, S.: Phys. Rev. **131**, 440 (1963)
18. Tricomi, F.G.: Integral Equations. Dover, New York, (1985) (chapter II)
19. Oller, J.A., Oset, E.: Phys. Rev. D **60**, 074023 (1999)
20. Chew, G.F., Mandelstam, S.: Phys. Rev. **119**, 467 (1960)
21. Castillejo, L., Dalitz, R.H., Dyson, F.J.: Phys. Rev. **101**, 453 (1956)
22. Chew, G.F., Frautschi, S.C.: Phys. Rev. **124**, 264 (1961)
23. Adler, S.L.: Phys. Rev. **137**, B1022 (1965)
24. Oller, J.A.: Phys. Lett. B **477**, 187 (2000)
25. Chanowitz, M.S., Golden, M., Georgi, H.: Phys. Rev. D **36**, 1490 (1987)
26. Bjorken, J.D.: Phys. Rev. Lett. **4**, 473 (1960)
27. Weinberg, S.: The Quantum Field Theory of Fields. Modern Applications, vol. II. Cambridge University Press, New York (2005)
28. Leutwyler, H.: Ann. Phys. **235**, 165 (1994)
29. Pich, A.: Rept. Prog. Phys. **58**, 563 (1995)
30. Bernard, V., Kaiser, N., Meißner, U.G.: Int. J. Mod. Phys. E **4**, 193 (1995)

© The Author(s), under exclusive licence to Springer Nature Switzerland AG 2019

J. A. Oller, *A Brief Introduction to Dispersion Relations*, SpringerBriefs in Physics,
https://doi.org/10.1007/978-3-030-13582-9

31. Ecker, G.: Prog. Part. Nucl. Phys. **35**, 1 (1995)
32. Oller, J.A., Oset, E., Pelaez, J.R.: Phys. Rev. D **59**, 074001 (1999), ((E) *ibid* **60**,099906 (1999); (E) *ibid* **75**, 099903 (2007))
33. Oller, J.A., Meißner, U.G.: Phys. Lett. B **500**, 263 (2001)
34. Tanabashi, M., et al.: Partice data group. Phys. Rev. D **98**, 030001 (2018)
35. Peláez, J.R.: Phys. Rept. **658**, 1 (2016)
36. Bando, M., Kugo, T., Uehara, S., Yamawaki, K., Yanagida, T.: Phys. Rev. Lett. **54**, 1215 (1985)
37. Bando, M., Kugo, T., Yamawaki, K.: Phys. Rept. **164**, 217 (1988)
38. Meißner, U.G.: Phys. Rept. **161**, 213 (1988)
39. Ecker, G., Gasser, J., Pich, A., de Rafael, E.: Nucl. Phys. B **321**, 311 (1989)
40. Kawarabayashi, K., Suzuki, M.: Phys. Rev. Lett. **16**, 255(1966); Rizuddin and Fayyazuddin, Phys. Rev. **147**, 1071, (1966)
41. Kaiser, N.: Eur. Phys. J. A **3**, 307 (1998)
42. Bernard, V., Kaiser, N., Meißner, U.G.: Nucl. Phys. B **364**, 283 (1991)
43. Manohar, A.V.: Large N_C QCD, hep-ph/9802419. Contribution to the proceedings of the conference Probing the standard model of particle interactions. Proceedings, Summer School in Theoretical Physics, NATO Advanced Study Institute, 68th session, Les Houches, France, July 28–September 5 (1997)
44. Briceno, R.A., Dudek, J.J., Edwards, R.G., Wilson, D.J.: Phys. Rev. Lett. **118**, 022002 (2017)
45. Nieves, J., Ruiz Arriola, E.: Phys. Lett. B **455**, 30 (1999)
46. Lichtenberg, D.B.: Unitary Symmetry and Elementary Particles. Academic Press, New York (1950)
47. Jido, D., Oller, J.A., Oset, E., Ramos, A., Meißner, U.G.: Nucl. Phys. A **725**, 181 (2003)
48. Oller, J.A.: Phys. Lett. B **477**, 187 (2000)
49. Cornwall, J.M., Levin, D.N., Tiktopoulos, G.: Phys. Rev. D **10**, 1145 (1974); Vayonakis, C.E.: Lett. Nuovo. Cim. **17**, 383 (1976); Lee, B.W., Quigg, C., Thacker, H.: Phys. Rev. D **16**, 1519 (1977); Chanowitz, M.S., Gaillard, M.K.: Nucl. Phys. B **261**, 379 (1985); Yao, Y.P., Yuan, C.P.: Phys. Rev. D **38**, 2237 (1988)
50. Weinberg, S.: Phys. A **96**, 327 (1979)
51. Gasser, J., Leutwyler, H.: Ann. Phys. **158**, 142 (1984)
52. Dobado, A.: Phys. Lett. B **237**, 457 (1990)
53. Pelaez, J.R.: Phys. Rev. D **55**, 4193 (1997)
54. Truong, T.N.: Phys. Rev. Lett. **61**, 2526 (1988); *ibid* **67**, 2260 (1991)
55. Dobado, A., Pelaez, J.R.: Phys. Z. C **57**, 501 (1993)
56. Dobado, A., Herrero, M.J., Truong, T.N.: Phys. Lett. B **235**, 129 (1990)
57. Dobado, A., Pelaez, J.R.: Phys. Rev. D **56**, 3057 (1997)
58. Oller, J.A., Oset, E., Pelaez, J.R.: Phys. Rev. Lett. **80**, 3452 (1998)
59. Gomez Nicola, A., Pelaez, J.R., Rios, G.: Phys. Rev. D **77**, 056006 (2008)
60. Nieves, J., Pavon Valderrama, M., Ruiz Arriola, E.: Phys. Rev. D **65**, 036002 (2002)
61. Gomez Nicola, A., Pelaez, J.R.: Phys. Rev. D **65**, 054009 (2002)
62. Nebreda, J., Pelaez, J.R.: Phys. Rev. D **81**, 054035 (2010)
63. Gomez Nicola, A., Pelaez, J.R.: Phys. Rev. D **62**, 017502 (2000)
64. Guo, Z.H., Oller, J.A., Ros, G.: Phys. Rev. C **89**, 014002 (2014)
65. Albaladejo, M., Oller, J.A.: Phys. Rev. C **84**, 054009 (2011); *ibid* **86**, 034005 (2012)
66. Oller, J.A.: Phys. Rev. C **93**, 024002 (2016)
67. Kaiser, N., Brockmann, R., Weise, W.: Nucl. Phys. A **625**, 758 (1997)
68. Molina, R., Nicmorus, D., Oset, E.: Phys. Rev. D **78**, 114018 (2008)
69. Du, M.L., Gülmez, D., Guo, F.K., Meißner, U.G., Wang, Q.: Eur. Phys. J. C **78**, 988 (2018)
70. Geng, L.S., Oset, E.: Phys. Rev. D **79**, 074009 (2009)
71. Borasoy, B., Nissler, R., Weise, W.: Phys. Rev. Lett. **96**, 199201 (2006)
72. Oller, J.A., Prades, J., Verbeni, M.: Phys. Rev. Lett. **96**, 199202 (2006); *ibid* **95**, 172502 (2005)
73. Abarbanel, H.D., Goldberger, M.L.: Phys. Rev. **165**, 1594 (1968)
74. Morgan, D., Pennington, M.R.: Z. Phys. C **37**, 431 (1988); (E) *ibid* **39**, 590 (1988)

75. Babelon, O., Basdevant, J.-L., Caillerie, D., Mennessier, G.: Nucl. Phys. B **113**, 445 (1976)
76. Jamin, M., Oller, J.A., Pich, A.: Nucl. Phys. B **622**, 279 (2002)
77. Brodsky, S.J., Farrar, G.R.: Phys. Rev. Lett. **31**, 1153 (1973)
78. Matveev, V.A., Muradian, R.M., Tavkhelidze, A.N.: Lett. Nuovo Cim. **7**, 719 (1973)
79. Brodsky, S.J., Llanes-Estrada, F.J.: arXiv:1810.08772 [hep-ph] (and references therein)
80. Levinson, N.: Kgl. Danske Videnskab. Selskab, Mat. Fys. Medd. **25**, No. 9 (1947)
81. Guo, Z.H., Oller, J.A., Ruiz de Elvira, J.: Phys. Rev. D **86**, 054006 (2012)
82. Oller, J.A., Roca, L.: Phys. Lett. B **651**, 139 (2007)
83. Donoghue, J.F., Gasser, J., Leutwyler, H.: Nucl. Phys. B **343**, 341 (1990)
84. Moussallam, B.: Eur. Phys. J. C **14**, 111 (2000)
85. Yndurain, F.J.: Phys. Lett. B **612**, 245 (2005)
86. Yndurain, F.J.: Phys. Lett. B **578**, 99 (2004); (E) *ibid* **586**, 439 (2004); hep-ph/0510317
87. Pennington, M.R.: Phys. Rev. Lett. **97**, 011601 (2006)
88. Oller, J.A., Roca, L., Schat, C.: Phys. Lett. B **659**, 201 (2008); Oller, J.A., Roca, L.: Eur. Phys. J. A **37**, 15 (2008)
89. Bijnens, J., Cornet, F.: Nucl. Phys. B **296**, 557 (1988)
90. Donoghue, J.F., Holstein, B.R., Lin, Y.C.: Phys. Rev. D **37**, 2423 (1988)
91. Muskhelishvili, W.I.: Singular Integral Equations. North-Holland, Amsterdam (1958)
92. Warnock, R.L.: Nuovo Cimento **50**, 894 (1967); (E) *ibid* **52A**, 637 (1967)
93. Jamin, M., Oller, J.A., Pich, A.: Nucl. Phys. B **587**, 331 (2000)
94. Gasser, J., Leutwyler, H.: Nucl. Phys. B **250**, 465 (1985); *ibid*, 517 (1985); *ibid*, 539 (1985)
95. Aston, D., et al.: Nucl. Phys. B **296**, 493 (1988)
96. Kang, X.W., Guo, Z.H., Oller, J.A.: Phys. Rev. D **94**, 014012 (2016)
97. Flatté, S.M.: Phys. Lett. B **63**, 224 (1976)
98. Kang, X.W., Oller, J.A.: Eur. Phys. J. C **77**, 399 (2017)
99. Guo, Z.H., Oller, J.A.: Phys. Rev. D **93**, 096001 (2016)
100. Weinberg, S.: Phys. Rev. **130**, 776 (1963)
101. Baru, V., Hanhart, C., Kalashnikova, Y.S., Kudryavtsev, A.E., Nefediev, A.V.: Eur. Phys. J. A **44**, 93 (2010); Bogdanova, L.N., Hale, G.M., Markushin, V.E.: Phys. Rev. C **44**, 1289 (1991)
102. Guo, F.K., Hanhart, C., Meißner, U.G., Wang, Q., Zhao, Q., Zou, B.S.: Rev. Mod. Phys. **90**, 015004 (2018)
103. Valderrama, M.P.: Phys. Rev. D **85**, 114037 (2012)
104. Weinberg, S.: Phys. Rev. **137**, B672 (1965)
105. Baru, V., Haidenbauer, J., Hanhart, C., Kalashnikova, Y., Kudryavtsev, A.E.: Phys. Lett. B **586**, 53 (2004)
106. Hyodo, T., Jido, D., Hosaka, A.: Phys. Rev. C **85**, 015201 (2012)
107. Aceti, F., Oset, E.: Phys. Rev. D **86**, 014012 (2012)
108. Sekihara, T.: Phys. Rev. C **95**, 025206 (2017)
109. Oller, J.A., Lacour, A., Meißner, U.-G.: J. Phys. G **37**, 015106 (2010)
110. Fetter, A.L., Walecka, J.D.: Quantum Theory of the Many-Particle Systems. Dover Publications Inc, Mineola, New York (1971)
111. Bernard, V., Kaiser, N., Meißner, U.-G.: Int. J. Mod. Phys. E **4**, 193 (1995)
112. Jamin, M., Oller, J.A., Pich, A.: hep-ph/0110193, eprint version of Ref. [93]
113. Gradshteyn, I.S., Ryzhik, I.M.: Tables of Integrals, Series, and Products. Academic Press, London (1980)
114. Press, W.H., et al.: Numerical Recipes. Cambridge University Press, Cambridge (1992)

Printed in the United States
By Bookmasters